Ethics of the Use of Human Subjects in Research

Ethics of the Use of Human Subjects in Research

Practical Guide

Adil E. Shamoo, Ph.D.
University of Maryland
School of Medicine
108 North Greene Street
Baltimore, Maryland 21201-1503

Felix A. Khin-Maung-Gyi, Pharm.D., M.B.A.
Chesapeake Research Review, Inc.
The Chesapeake Building
9017 Red Branch Road, Suite 100
Columbia, Maryland 21045

London and New York

Library of Congress Cataloging-in-Publication Data
Shamoo, Adil E.
Ethics of the use of human subjects in research / Adil E. Shamoo, Felix A. Khin-Maung-Gyi.
p. cm
Includes bibliographical references and index.
ISBN 0-8153-4073-7 (pbk.)
1. Human experimentation in medicine—Moral and ethical aspects. 2. Medicine—Research—Moral and ethical aspects. 3. Medical ethics. I. Khin-Maung-Gyi, Felix A. II. Title.

R853.H8 S535 2002
174'.28—dc21

2001059233

Published by Garland Science Publishing, a member of the Taylor & Francis Group, 29 West 35th Street, New York, NY 10001-2299.

Printed and bound in Great Britain by T.J. International Ltd, Padstow, Cornwall

15 14 13 12 11 10 9 8 7 6 5 4 3 2 1

To Abe, Zach, and Jessica,
and Rebecca

Contents

Preface

In the past 15 years, media headlines about research misconduct in American universities have focused public attention on the dramatic ethical problems that can arise during the conduct of research. In some instances, investigators have been accused, and occasionally found guilty, of the falsification, fabrication, and plagiarism of data or the abuse of human subjects in research protocols. A recent example is the death of an 18 year-old research participant in a gene transfer experiment.

There is widespread concern that public confidence in the scientific research establishment has become undermined. In the current atmosphere of added accountability, scientific research enterprise as a whole is now under increased scrutiny, not only by the media but by Congress and the public at large.

There have been congressional hearings, proposed legislative remedies, and regulations and guidance from federal oversight agencies to manage conflicts of interest. In addition, there have also been: institution-specific policies to deal with research misconduct; national conferences on a variety of related themes; the creation of a national Commission on Research Integrity; continuing deliberations of the National Bioethics Advisory Commission; the recent creation of the National Human Research Protections Advisory Commission; and a proliferation of university-based research ethics courses, as well as less formal academic forums for discussing issues with ethical impact.

While, currently, university courses in research ethics are not mandated for those health care and related professionals interested in pursuing careers in research, it is clear that their number is increasing especially on the use of human subjects in research protocols. For some years, all trainees (students, fellows, and others) who are receiving grants from the National Institutes of Health have been required to be exposed to issues related to research ethics. In addition, the Commission on Research Integrity has recommended that all participants in federally funded research should receive training in responsible conduct in research. Institutional Review Boards (IRBs), the lynchpin for the protection of human subjects in research, have also been a topic of public and regulatory debate, which has included suggestions for revamping the entire IRB system, most recently proposed by the National Bioethics Advisory Commission.

However, when it comes to human subject protection, as Dr Greg Koski, the Director of the Office of Human Research Protections, has repeatedly suggested, concern for participant safety is the responsibility of all who are involved in the conduct of clinical trials: sponsors, sponsors' representatives, research sites, and health care administrators, as well as IRBs.

There is a growing number of workshops, mini-courses, and training sessions in the proper use of human subjects in research. Furthermore, there are several drafts of proposed pieces of legislation, regulations, and guidance instituting the training and education of personnel involved in human research. There is very little disagreement among all parties about the need for training and education in this area.

As anticipated, on May 23 2000, the Secretary of the Department of Health and Human Services, Donna Shalala, announced that her department will seek new regulatory measures to enforce the protection of human subjects enrolled in research. She proposed fines of up to 1 million dollars. More importantly, Secretary Shalala announced on June 6, 2000, that as of that day the Department is requiring all clinical investigators to undergo training and education in the ethics of the use of human subjects in research. Furthermore, a new Office for Human Research Protection (OHRP) was announced to replace the former Office of Protection from Research Risks. The OHRP is located within the Secretary's office.

The academic community has responded positively to the need for learning materials and strategies providing support for training programs or for other experiences related to the ethics of the use of human subjects in research. This work, entitled *Ethics of the Use of Human Subjects in Research (Practical Guide),* is intended to fill that need. It will present a practical introduction to the ethical issues at stake. Beginning with a general discussion of research ethics, it spans topics such as the development of a protocol for ethical decision making through how to obtain an IRB approval, with an emphasis on the ethical factors underpinning the IRB review process.

Ethics of the Use of Human Subjects in Research (Practical Guide) is primarily intended for IRB members, researchers whose protocols involve human subjects and their staff, and students in clinical research. It can serve as a basis for those audiences' training or may supplement already established courses on research ethics. Research mentors who choose to approach the teaching of research ethics on an informal basis will also find this book helpful.

We apologize for any errors in this first edition. Please feel free to send comments and suggestions to: Adil E Shamoo, PhD, Department of Biochemistry and Molecular Biology, University of Maryland School of Medicine, 108 North Greene Street, Baltimore, Maryland 21201-1503; e-mail: ashamoo@umaryland.edu (AES only).

About the Authors

Adil E Shamoo, PhD

Dr Shamoo is a research scientist with over 30 years' experience in the laboratory, who has published over 200 papers in the fields of biophysics, biochemistry, ethics, science and public policy. He has authored, co-authored, edited, or co-edited 13 books. He is a professor and former chairman of the Department of Biochemistry and Molecular Biology, a member of the graduate faculty of Applied Professional Ethics, and is affiliated with the Center for Biomedical Ethics at the University of Maryland, Baltimore. He holds certificates of attendance on the Patent Review Course and a course on Regulatory Issues in Biotechnology. In 2001, Dr Shamoo received the certification of IRB Professionals (CIP) from the Council for Certification of IRB Professionals, of the Applied Research Ethics National Association (ARENA). Dr Shamoo consults on ethics, science, responsible conduct of research, and human research protections. He has chaired eight international conferences on ethics in research and public policy. The last three, in 1995, 1998, and 2000, were on the use of human subjects in research. Dr Shamoo has testified on this issue before congressional committees and the National Bioethics Advisory Commission. He served from 1995 to 1998 on the Maryland Attorney General's Task Force to propose legislation to protect decisionally impaired human subjects in research. When serving on the Board of Directors of the National Alliance for the Mentally Ill, he was the principal author of their 1997 Code of Ethics for enrolling mentally ill people in research. In addition, Dr Shamoo currently serves as a member of the Board of Directors of the Friends Research Institute, a national research and philanthropic organization. In the past, he has served, sometimes as President, on numerous boards, and on councils at local, state, and national level.

Dr Shamoo founded in 1988 and is the Editor-in-Chief of the journal *Accountability in Research*. He has given talks on this topic worldwide. He has held visiting professorships in institutions such as the Institute for

Political Studies in Paris, France, and at East Carolina University. His current interests in the areas of ethics, science, and public policy are the development of good research practices; ethics and public policy; objectivity and conflict of interest; and the use of humans in research, especially persons with mental illness. Since 1991 Dr. Shamoo has taught graduate students on the responsible conduct of research and held workshops on ethics in research with human subjects. He has been cited in the national print media, and has frequently featured in local and national radio and television programmes.

In December 2000 Dr Shamoo was appointed to the National Human Research Protections Advisory Committee (NHRPAC).

Felix A Khin-Maung-Gyi, PharmD, MBA, CIP

For more than 15 years Dr Khin-Maung-Gyi has been a consultant and has been active in the field of professional development as well as managing research units and projects. He has worked internationally, both traveling and living abroad. His roles have included being a clinical scientist and a project and program leader in both private and academic sectors. He has also restructured and managed an academic medical center Institutional Review Board (IRB). In addition to research project management and coordination, his other areas of expertise include human research subject protection, pharmacovigilance, and inter-sector partnerships in industry, government, academia, and non-profit-making organizations. His professional appointments have included IRB Forum Chair and Task Force Member, Principal Investigator Certification, Association of Clinical Research Professionals, and Member of the Council for Certification of IRB/IEC Professionals of the Applied Research Ethics National Association. He is also on the Editorial Board of *Clinical Trials Advisor*. Recently, Dr Gyi became the first, and is the only, independent IRB representative to be certified as an IRB professional by the Council for Certification of IRB/IEC Professionals, Applied Research Ethics National Association. Previously, he has worked for pharmaceutical (both domestic and international), medical, and professional services, and health care organizations. Throughout his professional career, Dr Gyi has also served as a university faculty member, including, until recently, at an academic medical center. He is a frequent speaker, moderator, and panelist at forums ranging from national professional conferences to Investigator Meetings.

Dr Gyi is a co-author of the following forthcoming or recent publications: "Electronic data capture issues for IRBs: administrative and regulatory," in *DIA Journal; Ethics of the Use of Human Subjects in Research;*

and "Ethical and confidentiality-based issues in patient recruitment today," a chapter in *A Guide to Patient Recruitment: Today's Best Practices and Proven Strategies*. He was also an expert panel member, contributing to the new volume of *The Complete Guide to Informed Consent in Clinical Trials*.

He received a Doctor of Pharmacy degree from Duquesne University, and an MBA (Executive Program) from Loyola (Maryland) College. He has completed clinical pharmacy residencies in both pediatric and adult medicine and carried out postgraduate studies in drug development, regulations, and medical research. His undergraduate work in pharmacy concentrated on microbiology.

In 2000 the company he founded in 1993 gained the Blue Chip Enterprise Award (sponsored by the US Chamber of Commerce). In the previous year, the company was named in both the Fast 50 (regional) and Fast 500 (national) lists, a program sponsored by Deloitte and Touche for companies displaying growth and outstanding accomplishment.

Acknowledgements

We are grateful for the invaluable comments and suggestions of Dr Mathew Whalen of Chesapeake Research Review, Inc., received throughout the preparation of this monograph. We would like to thank Ms Terry Gyi for her comments and suggestions on the draft chapters, and also Zoltan Annau, PhD, former professor at Johns Hopkins School of Hygiene.

List of Abbreviations

ADR	adverse drug reaction
AE	adverse event
AMC	academic medical center
CFR	Code of Federal Regulations
CRO	clinical research organization
DHHS	Department of Health and Human Services
DSMB	Data Safety Monitoring Board
FDA	Food and Drug Administration
FWA	Federal Wide Assurance
ICH	International Council on Harmonization
IND	investigational new drug
IRB	institutional review board
NBAC	National Bioethics Advisory Commission
NHRPAC	National Human Research Protections Advisory Committee
NIH	National Institutes of Health
OHRP	Office for Human Research Protection
OPRR	Office for the Protection from Research Risks
PPRA	Protection of Pupil Rights Amendment
SSP	Special Standing Panel
US	United States (adj)
USA	United States of America (noun)

Chapter 1

Introduction to Research Ethics

The use of human subjects in research is critical to developing basic and applied knowledge in the biomedical and social sciences. In clinical research, experiments on human beings are essential to testing and developing new treatments to combat disease and promote health.

Human subject involvement through clinical trials has been expanding for well over a decade, as have ethical concerns. The greater research effort has been accompanied by increasing concern with ethical issues owing to a variety of factors:

- *An aging population in the developed world and a longer life span in some developing countries*
- *Public policy focus on health care and health care coverage in North America, the European Community, and Japan, and in the developed world in particular*
- *Public awareness of research risks, nationally and internationally, as publicized throughout the media, especially the interrelationships between the developed and the developing worlds*
- *Increasing US federal government funding for health related research*
- *Increased private funding of clinical research, which, for the last 15 years or so in the USA, has outstripped federal government funding*

The use of human research subjects plays a key role in promoting disease treatment and prevention, public health, and social well-being. However, the vulnerability of many participants to potential abuses, especially those with significant diseases (such as chronic conditions or those that impact a person's decision-making capacities), raises considerable challenges from business, government regulatory, and ethical perspectives.

In this chapter, relevant ethical theories are briefly overviewed, including virtue ethics, Kantianism, and utilitarianism. From these theories, four principles are derived: respect for persons, beneficence (do good), nonmaleficence (do no harm), and justice.

Compliance with ethical principles is not only important for society, it also contributes to maintaining the public confidence that is so essential to the continuance of the development of new and improved treatment strategies. Researchers, as leaders in society, are expected to promote ethical values as well as to provide vision, initiative, and realism.

Learning Objectives

1. The necessity of using human subjects in research
2. The unprecedented growth of research in the past few decades
3. The Nuremberg Code (1947) as a cornerstone of ethics of the use of human subjects in research
4. The basic ethics theories
5. The most commonly used ethical principles
6. The characteristics of a researcher as a leader

Modern science is based on the continuous acquisition of new knowledge. This now occurs primarily through systematic and organized research undertakings involving multiple investigators and, very frequently, large operations.

The importance of research in advances in national defense, the economy, education, the environment, and health care has been well documented for at least the past 100 years. These advances create new technologies and innovations that keep our economies competitive, and new tools and drugs for patient diagnosis and therapy. For example, the total expenditure in the USA for research and development exceeds $170 billion annually, with federal government funds accounting for less than half of this.

The significant growth of public funding for research in the USA started after World War II (STAE, 1989; Shamoo, 1989). The fruits of research were recognized when resulting knowledge made an important and crucial impact on the outcome of the War. Through grants and contracts, the US Government began to fund university scientists to conduct research in numerous fields such as agriculture, medicine, and science in general. This was a new age for biomedical science research and knowledge, emerging from an unprecedented growth in work largely conducted by university scientists and at the federal campus of the National Institutes of Health (NIH) in Bethesda, Maryland. Since the mid-1980s in particular, industry as well as government have increased dramatically their funding for biological research (Beardsley, 1994).

Research is always carried out in a social and political context, whether initiated by individuals or by collaborations. Thus, societal priorities have a direct impact on what kind of research is funded. Even though researchers strive for objectivity, societal ethical issues are intertwined with the purpose and process of their research. They need to earn and then maintain public trust in order to gain support.

There have been high-profile incidences of research aberrations that have shaped modern discussions of research ethics. Examples include concerns over the use of pesticides, air pollution, radiation, and the human atrocities before and during World War II, including, in the USA, the Tuskegee experiments, and, during the war, experimentation by German and Japanese scientists. The rapid growth of industry, especially the biomedical industry after World War II and more recently the growth of biotechnology, have added greater complexities to the conduct of research, particularly with respect to ethics. In turn, even though the industrial funding for research has been welcomed by most researchers, it has added commercial dimensions to ethical issues for consideration.

Intensified research activity in the USA and internationally have also resulted in significant increases in the use of animals and humans as experimental subjects. The use of people in research is necessitated if new drugs are ever to be used by humans. It is critically important for this to be understood by all the parties involved (i.e. patients, families, researchers, and industrial concerns). If public confidence fails, the use of human subjects in research will decrease and drug development will suffer. If new drugs do not reach the market, desperate patients may use untested and unproven chemicals and drugs to the detriment of their health.

All those involved in clinical research must come to the conclusion that the use of human subjects can and must be conducted in an ethical manner in order to serve the health and well-being of our society.

During World War II, Nazi "researchers" used prisoners – mostly Jews, and mentally retarded and mentally ill people – in experiments that shocked the conscience of the world as they came to light during the Nuremberg trials. The Nuremberg military tribunal issued a strong condemnation of such barbaric behavior towards fellow human beings and issued a code of conduct for research with humans, now known as the Nuremberg Code (ACHRE, 1995). Unfortunately, while paling in comparison with the Nazi scale, human research subject abuse was discovered in the USA prior to and, more disturbingly, after World War II.

The Nuremberg Code identified crucial elements for human subjects' participation in research. The first and most enduring principle is that "The voluntary consent of the human subject is absolutely essential." The Code further emphasizes the requirement that the subject must be given sufficient knowledge, be free of duress and coercion, and comprehend the associated risks and benefits.

In 1964, the Helsinki Declaration, adopted by the World Medical Assembly meeting in Helsinki, Finland, followed the Nuremberg Code. It has since been revised five times (for general references, see Beauchamp and Childress, 1994; ACHRE, 1995; Veatch, 1989). The Helsinki Declaration provides more detail on regulating the protection of human research subjects and distinguishes between therapeutic and nontherapeutic research.

ETHICAL PRINCIPLES

In our society, there are at least two distinct meanings for the term "ethics". In the first sense of the term, ethics refers to principles of conduct that distinguish between right and wrong actions. These principles may be embodied in a particular code, creed, or law, such as the Nuremberg Code, or they may exist as social conventions or norms (or morals). Beauchamp and Childress (1994) link ethics and morality by asserting:

> "The term morality refers to social conventions about right and wrong human conduct that are so widely shared that they form a stable (although usually incomplete) communal consensus, whereas ethics is a general term referring to both morality and ethical theory."

In the second sense of the term, "ethics" refers to the scholarly study of right and wrong, which has been traditionally addressed by humanistic disciplines such as philosophy, religious studies, law, and social science. There are a variety of ways of approaching the study of ethics:

- Normative ethics, which covers general theories and principles of conduct
- Theoretical ethics (or meta-ethics), which concerns the meaning and justification of concepts and theories of ethics
- Applied ethics, which includes ethical dilemmas and problems as they arise in particular situations, such as medicine or clinical research
- Descriptive ethics, which is a scientific attempt to describe and explain the ethical values and beliefs held by different people in society and around the world (Beauchamp and Childress, 1994)

Codes of ethics evolved from secular and religious underpinnings. The earliest foundation of medical ethics began with the Hippocratic oath over 2400 years ago and Hippocrates' disciples: "first do no harm." The oath emphasized that the physician should do his or her best according to the physician's "ability and judgment" and preserve the patient's confidentiality (Veatch, 1989). It has been criticized for promoting paternalism, which, in medicine, implies that patients surrender their judgment (i.e. autonomy) to the physician.

More recent codes of ethics have evolved, beginning with that of the American Medical Association, nearly 160 years ago, which has undergone many revisions since (Veatch, 1989). Paternalism was also a hallmark of this code, as of others of the time. Subsequently, most medical professional groups developed their own codes.

Secular liberal philosophies added to our understanding of medical ethics. Historical figures such as Locke, Jefferson, and Madison, and documents such as the US Bill of Rights all contributed to the prominence of the individual's rights (i.e. autonomy) in ethical thinking. The Bill of Rights was written to protect the minority from the tyranny of the majority. It emphasized that, in order to live in a civil society, certain minimal rights must be protected regardless of the majority views.

The modern terminology associated with ethical principles in research are those associated with ethics in general and are derived from philosophical underpinnings dating back to Socrates. Ethical dilemmas come into focus when people have to make a personal decision for themselves or their loved ones. Researchers may face ethical challenges, not only in their laboratory when it comes to the integrity of their data but also when a clinician/researcher has to test a new drug therapy that may carry risks more, equal to, or less than the standard therapy in terms of being either safe or effective.

Among moral theories (Beauchamp and Childress, 1994), the most relevant are the following:

1. *Virtue-based ethics* Following the traditions of Plato and Aristotle, emphasizing virtue and the good life, virtue ethics focuses on the character of the actor rather than the act itself. Usually it is hard to separate the act from the actor. The actor may be good but the consequence of the action may be bad. Therefore, ethical actors or agents are supposed to act in an ethical manner (also known as character ethics). These character values are important for a moral society. Some of the most often mentioned virtues include compassion, veracity (truth telling), trustworthiness, integrity, courage, humility, temperance, justice, and confidentiality. In the context of research on human subjects, the character of the researcher is important regardless of whether the researcher's acts are open to public scrutiny or not, because society must trust that researchers have a good enough character to enable them to do the right thing even when their actions are not, or may never be, scrutinized.

2. *Kantianism* This theory was enunciated by the German enlightenment philosopher Immanuel Kant (1724–1804). He held that the rightness or wrongness of an action resides in the act itself rather than in its consequences. Respect for persons (autonomy) and universality ("what if everybody did this?"), better known as the categorical imperative, are two principles that are derived from this moral theory. Moral theories that focus on actions themselves rather than consequences are called deontological theories.
3. *Utilitarianism* This theory holds that one should perform the action that promotes the greatest balance of positive (as opposed to negative) consequences for the greatest number of people. Utilitarianism was first enunciated by Jeremy Bentham (1748–1832) and John Stewart Mill (1806–1873). Bentham and Mill used happiness to measure utility; other writers have addressed different measures of utility, such as the satisfaction of preferences, and economic costs and benefits. Risk–benefit (cost–benefit) analysis, which is widely used in health care and research with human subjects, is derived from this ethical principle. In risk–benefit analysis, we are weighing the social benefits from an act, comparing them with the risk to the individual. However, an unchecked utilitarian (cost–benefit) analysis approach can lead to violations of other principles of ethics.

   Additional ethical principles have evolved throughout history. Among them is that of natural law, dating from the ancient Greeks, especially Aristotle. Among the naturally good things are life, health, and happiness. Among the naturally bad things are death, disease, and suffering. Therefore, our ethical behavior requires the promotion of naturally good things in contrast to naturally bad things.

   Another ethical principle is that of natural rights, developed by the British philosopher, physician, and theologian John Locke (1632–1704). Locke promoted that all people should have access to the basic rights of life, liberty, and property. The principle embodied in the US declarations and constitutions are Lockean.

Modern ethical principles are derived, in large part, from these three moral theories and can be seen at work in contemporary applied ethics discussions, including, for example, the work of the National Commission for the Protection of Human Subjects of Biomedical and Behavioral Research (known as the National

Commission) which issued a report called the Belmont Report (National Commission, 1979). This articulated the following four principles, which reflect the history of moral theory, now specifically applied to clinical research:

1. Respect for the person (autonomy)
2. Beneficence (do good)
3. Nonmaleficence (do no harm)
4. Justice (fairness and utility)

Respect for the Person (Autonomy)

This important moral principle was the cornerstone of Immanuel Kant's restatement of the Aristotelian, and Judeo–Christian, saying, "Treat others as you want them to treat you." In other words, human beings should never be treated as only a means to another end; they are ends in themselves and have inherent value or worth. This principle is also the basis of numerous other moral precepts such as confidentiality, freedom of choice, accountability, avoidance of conflict of interest, and informed consent. Embedded in these precepts is the ability of individuals to be autonomous in making their decisions. The lack of fear or favor, the lack of duress, the requirements for comprehensibility, and providing sufficient knowledge before a decision is to be made are each related to augmenting and insuring respect for persons. Each person has intrinsic worth and has specific rights that other individuals must respect, including those who lack decisional capacity, and children, who lack complete autonomy to make decisions owing to their inability to weigh risks and benefits to themselves.

It is important to appreciate the relativity, in cultures, of these principles. Respect for the person, especially in the USA, is considered to be the first principle that should not be violated, except in extreme and rare circumstances. This is clearly the view of libertarians. There are others who consider that this principle is of no more weight than others; indeed, some would argue that circumstances determine which principle in a particular case is inviolate.

Respect for the person also requires that prospective research subjects should be given sufficient information regarding risks and benefits in order to weigh properly the prospect of enrollment in the research project. Furthermore, respect for the person requires that the enrollment of an individual into a research study should be voluntary, in other words, free from any duress, coercion, and undue influence. In current discussions, this raises the question:

"Is it appropriate that a patient's physician should ask the patient to enroll?" This is especially controversial if that physician is the patient's psychiatrist (NBAC, 1998).

Beneficence (Do Good)

This principle is intertwined with that of nonmaleficence and at times the two are lumped together. Primarily, it emphasizes enhancing kindness, charity, and the welfare of others. This utilitarian principle elevates our acts of charity to a moral obligation without committing harm to others in the process. It is not a wholly accepted principle because doing good to others becomes a demand of moral duty, whereas some contend it should be optional. The Hippocratic oath is a manifestation of this principle.

Beneficence requires that prospective research subjects should be given sufficient information to weigh risks against benefits. More importantly, the investigator, in the submission to the institutional review board (IRB), should discuss the risks and benefits with the subject. In addition, the research proposal should justify the use of human subjects in the first place, with documentation of the steps taken to minimize risks and to identify whether vulnerable populations are to be included, and, if so, if they are necessary to conduct the research.

Nonmaleficence (Do No Harm)

"Do no harm" is the primary admonition of the Hippocratic oath: "[A] Physician ought not to: inflict pain, suffering, and distress (physical or psychological), loss of freedom, disability and death. An individual should not deprive others from pleasures and happiness by restricting autonomy."

One ethicist (Frankena, 1963) prioritizes beneficence, first, by the duty not to inflict harm; secondly, to prevent harm; thirdly, to remove harm; and, finally, to promote good. Therefore, a researcher's obligation to research subjects to promote their good is less important than to prevent their harm. Invasive procedures such as biopsies are less acceptable than obtaining a sample of already available blood, for example.

Justice

Justice and fairness are terms used to connote equality of treatment of each person before an obligation or an authority. The

US constitutional principle of due process before the law is a clear manifestation of justice. In other words, if a process is established to punish or reward a US citizen, then the same process should be applied to all other citizens. This principle is thought to come from Aristotle: "Equals ought to be treated equally and unequals unequally." Justice requires a fair sharing of benefits and burdens, such as taxes, education, national services in case of war, and, some would say, equal access to health care.

Utilitarianism mixes justice with beneficence when invoking the concept of "the greatest good to the largest number of people." However, in western culture, as influenced by the Judeo–Christian code of ethics, protecting orphans, the disabled, and the weak is required. Bertrand Russell, the philosopher, went so far as to say that civilizations are measured by how they treat the most unfortunate among them.

In research with human subjects, justice requires that their selection should be fair to all individuals in that class. Therefore, those selected for research should reflect a fair sharing of burden and benefits as to their social, sexual, and ethnic characteristics. Social justice requires that subjects should not be selected because of race, decisional incapacity, or condition (for example, of being institutionalized). As one illustration, the testing of new cancer drugs should not be carried out on schizophrenic patients simply because they are more available to researchers.

ETHICS AND LEADERSHIP IN RESEARCH

Research undertaken by industry and academic institutions is a complex endeavor. It requires a great deal of funding as well as large numbers of personnel such as researchers, engineers, technicians, nurses, managers, and administrators, among others. The complex and diverse nature of research not only requires intellectual creativity and insight but, more importantly, ethical leadership. This is true for all who participate in research, not just research administrators.

The institutional culture of a research operation has a crucial impact on the quality and integrity of the outcome (i.e. the research data). Institutional culture within an operation impacts all aspects of the research data outcome. That is, this culture impacts the protection of human subjects, informed consent, respect for human beings, the avoidance of conflict of interest, and diligence in following all of the regulatory guidelines.

In addition to conducting himself or herself as a scientist with a keen and creative mind, each researcher is also a leader. Depending on his or her position in an operation, the impact of the researcher's leadership may be even more far-reaching than that of others. Therefore, the ethical stance of the leader will directly impact those led by him or her. For example, does the leader tolerate sloppy work, cutting corners, and a lack of concern for human subjects? He or she is the role model, especially in academic institutions where all students and postdoctoral fellows, the future leaders, are to be found. A leader inspires those working with him or her not just for greater creativity but also for a sense of duty to society and civilization. Regardless of the field, leaders have four primary characteristics (Koestenbaum, 1991; Gardner, 1995). In our view, ethics as a characteristic both embraces and clarifies the others:

1. *Ethics* In the context of leadership, ethics primarily emphasizes the traits of virtue ethics mentioned earlier in order to insure ethical leadership. These include, for example, integrity, truthfulness, trust, compassion, and a commitment to do no harm.
2. *Vision* Leaders need to have and maintain a broad perspective of the overall operation over which they preside. They provide solutions for the future, not just the short term. Vision is opposite to micromanagement. It requires great intellect for strategic thinking. George Bernard Shaw neatly captured part of this strategic thinking competency almost 100 years ago: "you see things; and you say 'why?' But I dream things that never were; and say, why not?"
3. *Initiative* This is the ability to *take* the initiative, to act, to take risks, to be determined and courageous. Leaders are willing to accept the consequence of their actions. They believe and practice accountability.
4. *Being realistic* Leaders' decisions are grounded in data and facts, not in illusions. This is probably what most researchers and leaders actually practice in their daily work. Researchers attempt to be objective, to make scientific decisions based on data and facts, while avoiding subjectivity. Research cannot survive for long if researchers are irrational. Mark Twain used to poke fun at those who did not always base their decisions on facts (namely the politicians of his time) when he wrote: "First get your facts, then you can distort them as much as you like."

> Researchers become so enmeshed with the cognitive/experimental process that leadership characteristics may become victims of neglect.

Leaders need to review regularly how they practice leadership in the context of these characteristics, especially ethics.

In this introductory chapter, we have laid out the general framework and principles of ethical conduct in research involving human subjects. These principles ought to be utilized not only because they are right but also because they maintain public confidence, which is essential for the research enterprise.

Quiz: Choose the Best Response

1. In modern science, especially in biomedical research, the acquisition of new knowledge is primarily gathered through:
   a. Case reports
   b. Individual investigators
   c. Private industry
   d. Systematic, organized, and large operations
   e. The NIH

2. The majority of NIH funding for research involving human subjects goes to:
   a. The pharmaceutical industry
   b. Intramural research on NIH campus
   c. Individual investigators and practitioners
   d. Hospitals
   e. Universities

3. Research on human subjects is important because:
   a. We cannot carry out such research on animals
   b. We cannot simulate all of the results on a computer
   c. Desperate patients may seek untested and unproven chemicals
   d. It will help to develop new and improved drugs
   e. All of the above

4. Codes of ethics in medical research in general evolved from the following, *except*:
   a. Abraham Lincoln
   b. AMA code of ethics
   c. Hippocratic oath
   d. Judeo–Christian ethics
   e. Lock and Kant

5. John Stewart Mill and Jeremy Bentham are identified with which ethical theories:
   a. Virtue-based ethics
   b. Utilitarianism
   c. Acts and not consequences are important
   d. Integrity
   e. Truth telling

6. The National Commission enunciated four broad principles of ethics:
   a. Respect for the person, beneficence, justice, and confidentiality
   b. Justice, nonmaleficence, beneficence, and autonomy
   c. Beneficence, justice, trustworthiness, and autonomy
   d. Respect for the person, beneficence, nomaleficence, and compassion

7. "Treat others as you want them to treat you" means:
   a. Human beings are means to an end
   b. A concept not related to informed consent
   c. A concept not related to conflict of interest
   d. Human beings are an end in themselves
   e. Comprehension is the sole determinant of informed consent

8. "Do good" in the context of medical ethics means:
   a. A requirement for an ethical society
   b. To enhance kindness and charity
   c. To prevent harm
   d. To enhance confidentiality
   e. All of the above

9. "Do no harm" may be interpreted as to require the researcher to:
   a. Not inflict harm
   b. Remove harm
   c. Promote good
   d. Alleviate suffering
   e. All of the above

10. The principle of justice requires the fair sharing of benefits and burdens such as taxes and national services. Is access to health care for all justified by the principle of justice?
    a. No, because not everyone can afford it
    b. Yes, because we are all created equal

c. Maybe, it depends on multiple factors
d. a and b are true
e. a and c are true

11. Leadership requires:
a. Vision
b. Initiative
c. Realism
d. Ethics
e. All of the above

References

Advisory Committee on Human Radiation Experiments (ACHRE) (1995) Final Report, stock number 061-000-00-848-9. Available from: Superintendent of Documents, US Government Printing Office, Washington, DC. Tel: (202) 512-1800; Fax: (202) 512-2250.

Beardsley T (1994) Big-time biology. *Sci. Am.* 271(5): 90–97.

Beauchamp TL & Childress JF (1994) Principles of Biomedical Ethics, 4th edn. New York: Oxford University Press.

Frankena WK (1963) Ethics, 2nd edn. Englewood Cliffs, New Jersey: Prentice Hall.

Gardner H (1995) Leading Minds – an Anatomy of Leadership. Boulder, Colorado: Basic Books.

Koestenbaum P (1991) Leadership – the Inner Side of Greatness. Oxford, UK: Jossey–Bass.

National Bioethics Advisory Commission (NBAC) (1998) Research Involving Persons with Mental Disorders That May Affect Decision-making Capacity, vol. I: Report and Recommendation. Rockville, Maryland: NBAC. Available at: http://www.bioethics.gov

National Commission for the Protection of Human Subjects of Biomedical and Behavioral Research (1979) Belmont Report: Ethical Principles and Guidelines for the Protection of Human Subjects of Research. Washington, DC: US Department of Health, Education, and Welfare.

Science and Technology in the Academic Enterprise (STAE) (1989) Status, Trends, and Issues – a Discussion Paper. The Government–University–Industry Research Roundtable, pp 1:2–72. Washington, DC: National Academy Press.

Shamoo AE (1989) Organizational structure and function of research and development. In: Principles of Research Data Audit (AE Shamoo ed.), pp 39–63. New York: Gordon and Breach.

Veatch RM (1989) Medical ethics: an introduction. In: Medical Ethics (RM Veatch ed.), pp 1–26. Boston, Massachusetts: Jones and Bartlett.

Chapter 2

Development of a Protocol for Ethical Decision Making

For nearly 100 years, there have been documented cases of human beings subjected to risk in research experiments. The atrocities carried out in Germany during World War II, as revealed during the Nuremberg trials, led to the formulation of guiding principles in what is now known as the Nuremberg Code. The Code declares that "voluntary consent is absolutely essential." Unfortunately, even in the USA, although on a lesser scale, there have been numerous cases of human subjects being used in high-risk experiments either without consent or by being deceived into giving consent. Examples of these are the Tuskegee syphilis studies and the Department of Energy radiation experiments. This chapter includes discussion of a method to be used for ethical decision making regarding the use of human subjects in research.

Learning Objectives

1. How the Nuremberg Code came into being
2. The most important principles of the Nuremberg Code
3. Why the National Commission was appointed
4. The key recommendation of the National Commission
5. How to analyze a case and reach an ethical decision

In 1989, Alex Capron wrote: "...the darkest moments in medical annals have involved abuses of human subjects." Some may consider this is an overstatement. Nevertheless, it can serve as a warning. Human history is filled with numerous stories illustrating human cruelty as well as strong stands on respecting others in the research environment. Humanity's drive for improved living conditions and the thirst for understanding nature led civilizations to seek ways to acquire new knowledge. Starting in the twelfth century, the pursuit of new knowledge in Europe gained momentum. This new understanding in methods of reasoning as well as content became the foundation of modern science. Some of these avenues of acquiring new knowledge were those that required testing new chemicals, drugs, and treatment modalities on human beings.

In 1900, in Prussia, nontherapeutic medical treatments were first prohibited. The use of consent by patients and the exclusion of children from experimentation were also affirmed (Capron, 1989). However, some researchers resorted to self-experimentation and experiments with family members in order to achieve their goals (Bean, 1977).

Immediately after World War II, the civilized world was shocked to learn of the atrocities of Nazi experimentation on unsuspecting disabled, retarded, and mentally ill people, and Jews (Proctor, 1988; Muller-Hill, 1988; Caplan, 1992). Some of these experiments entailed irradiation to induce sterility, high-altitude decompression, severe hypothermia, and numerous other experiments on prisoners and concentration camp residents. In imperial Japan, unwilling subjects, including prisoners, also underwent experimentation.

Over 100 years ago, Dr Walter Reed, the medical officer in post during the building of the Panama Canal, stood out as an example of ethical behavior at the time in his research experiments on yellow fever (Bean, 1977). Dr Reed prepared informed consent forms in both English and Spanish, making provision for significant monetary compensation to families if anyone died owing to the experiment. He also provided care for patients during their participation. None of his patients who contracted yellow fever died.

THE NUREMBERG CODE

After the Nuremberg trials of Nazi doctors, the tribunal judges of one of the trials included in their ruling certain principles of ethical conduct when using human beings in research. Owing to the strength of the words as well as their historical context, these pronouncements became known as the Nuremberg Code.

The most important and enduring principle was the first, declaring: "The voluntary consent of the human subject is absolutely essential." Some interpret the word "absolutely" to forbid any experiment regardless of the special circumstances. Others allow for the fact that the tribunal dealt mostly with adults who were forced to participate in barbaric experiments. Therefore, the tribunal's statement was made in that context. The tribunal was not addressing research with children or those who were decisionally impaired. Since then there has accumulated a vast amount of literature addressing the special circumstances and the ethical justification for softening the "absolute" standards. The Code also emphasizes respect for persons, which was to become a seminal principle in later statements on ethics. The Code further emphasizes that the consent of the individual must be based on sharing sufficient information with that person, and that the subject must be free of duress and coercion and must comprehend the information. The Nuremberg Code therefore represented a clear departure from the previous paternalistic stance taken towards research subjects by physicians and towards self-determination by the participants themselves (Veatch, 1989).

THE HELSINKI DECLARATION

At a meeting in Helsinki, Finland, in 1964, the World Medical Association issued important guidelines for the use of human subjects in research; this became known as the Helsinki Declaration. It reaffirmed the Nuremberg Code and distinguished between therapeutic and nontherapeutic research. This distinction lacked clarity and caused some confusion. The Declaration has since been revised six times (WMA, 2000).

HUMAN SUBJECT EXPERIMENTS IN THE USA BEFORE AND AFTER THE NUREMBERG CODE

A disturbing chapter in US history was the Tuskegee syphilis study sponsored by the Public Health Service. From 1932 to 1973, 400 poor African–American men living in Macon County, Alabama, were left to suffer from syphilis to observe the progress of the disease despite the fact that penicillin became available in the early

1940s (ACHRE, 1995). Prior to 1940, these patients were also denied the standard of care available at the time. They underwent numerous lumbar punctures, and were deceived and lied to regarding the nature of their illness. They were told that they had "bad blood." They were given some free meals and a $50 burial stipend. In 1997, President Clinton offered a public apology to the victims.

In 1954, a University of Chicago study conducted a secret recording of a jury deliberation in Wichita, Kansas. The judge and attorneys on the case were all aware of the recording and approved it. However, the jurors were not aware of it. The revelations of this case caused public outrage for undermining the jury system of justice and its confidential deliberations. Congress passed a law forbidding the recording of a jury's secret deliberations (Katz, 1972).

Another controversial and well-known case was the Yale University study known as the Milgram Obedience Study, carried out in 1963, the aim of which was to understand obedience and human authority. "Teachers" agreed to take part in an experiment to administer increasing intensities of electric shock to students giving wrong answers. In reality, the students were not receiving electric shocks because the teachers did not know that there was no electric charge in the machine. The students were screaming as if they had received a painful electric shock. Sixty percent of the teachers obeyed their orders and gave the electric shocks (Katz, 1972). This experiment raised the ethical issue of deception in research and under what restrictive conditions it should be allowed.

In the 1960s it became public knowledge that mentally retarded children at Willowbrook State School in New York had been injected with hepatitis virus. Even though the principal investigator obtained informed consent from parents, critics claimed that the parents were under the impression that their children were receiving a vaccine. In the mid-1960s live cancer cells were introduced into unsuspecting patients in a New York hospital. Hospital administration staff served as "whistle-blowers."

One of the most influential revelations came from Dr Henry Beecher (1966), a Harvard doctor, who chronicled in the *New England Journal of Medicine*, 22 examples of patients who had undergone risky experiments. The results of these experiments had been published in major medical journals and the research had been conducted by mainstream investigators. Beecher levied a strong indictment of these unethical practices.

Detailed accounts of these and other cases are chronicled in the report of the Advisory Committee on Human Radiation Experiments (ACHRE, 1995).

NATIONAL COMMISSIONS

After the Tuskegee revelations in the press, the first national commission was appointed in 1973, the National Commission for the Protection of Human Subjects of Biomedical and Behavioral Research. The Commission's report is simply known as the Belmont Report, and the Commission itself is known as the National Commission (National Commission, 1979). It identified the four core ethical principles that we reviewed in the previous chapter.

Another commission, the President's Commission for the Study of Ethical Problems in Medicine and Biomedical Behavioral Research, appointed in 1978, addressed broader issues than the 1973 National Commission, including genetics, health care of terminally ill people, and the use of mentally ill subjects in research (President's Commission, 1983).

In 1994 President Clinton appointed the Advisory Committee on Human Radiation Experiments (ACHRE, 1995), after reporter Eileen Welsome of the *Albuquerque Tribune* revealed that patients had been exposed to plutonium. Later it was discovered that, from 1944 to 1974, thousands of patients had been exposed to radiation in air during uranium mining and during medical experiments with radioactive substances on mentally retarded children. The Committee's report was nearly 1000 pages long. It consisted of well-documented cases of medical experiments in which human subjects had participated without their consent. It also documented the history of human subject use in research (ACHRE, 1995). The Committee recommended compensation and an apology to the victims, which President Clinton later delivered.

The most recent Commission was appointed in 1995 by President Clinton, the National Bioethics Advisory Commission (NBAC), and issued their last report in August 2001. It is important to note that the Commission's regular schedule was interrupted when the President requested them to the address the ethics of cloning. For 2 years, the NBAC addressed the issue of the use of mentally ill people in research. The Commission was responding to revelations, especially in the media, that mentally ill people had been exposed to high-risk experiments. The Commission made sweeping recommendation to strengthen protection for decisionally impaired people as research subjects (NBAC, 1998).

PROTOCOL FOR ETHICAL DECISION MAKING

From the four commonly cited ethical principles, moral duties become easier to identify. Individuals make moral choices daily without necessarily referring to any given set of ethical principles. Many different sources influence our moral values and decisions as we mature, including parents, relatives, friends, religion, education, the media, and social leadership. We also learn about moral values and decisions through our personal experiences of harm and benefit, respect and lack of respect, confidentiality and breach of confidentiality, honesty and dishonesty, and fairness and unfairness. We learn, as we grow up, how our actions affect other people and how the actions of others affect us. The moral dilemmas we often face in life help us learn how to make moral choices and decide to do the right thing. As a result, we develop a moral conscience, which causes us to feel guilt when we choose actions that we know to be wrong. We also develop a sense of pride or self-esteem when we choose the right actions for the right reasons.

However, we may also be confronted with choices in which there is no clear or obvious difference between right and wrong. In these situations, known as moral or ethical dilemmas, there may be several options that appear to be equally good or bad from a moral point of view. We must then decide how to solve these dilemmas. Many of the issues relating to the use of human subjects in research involve moral dilemmas. The following steps provide a useful guide for ethical decision making:

1. *Collect relevant information* In order to understand an issue one must collect the relevant (germane, essential) information. For example, it is vital to have sufficient information to evaluate the research protocol for risks and benefits, alternative treatments, and current medical practices for that particular protocol.
2. *Consider the different options* What are the different courses of action one may take? One cannot make a decision without knowing the options. In this step it is important to consider traditional as well as novel courses of

action since it is sometimes possible to develop unique solutions that minimize ethical conflict.

3. *Identify and prioritize the parties involved (i.e. stakeholders)* Examples of parties could be the human subjects, researchers, institutions, and society. It is also important to identify which party is primary, secondary, or tertiary in a particular case.
4. *Categorize the problems, questions, or issues within the four ethical principles* This will help to separate major ethical issues from auxiliary ones. For example, are we violating respect for persons? Recall that integrity, veracity, compassion, trustworthiness, and confidentiality are ethical characteristics of respect for persons. Therefore there are several subcategories that need to be identified. Is it fair to all prospective subjects? Is there any harm or potential harm to anyone? Is there weak or strong justification for harm or potential harm? Is there a need to be altruistic? Is the design of the experiment up to current scientific standards?
5. *Prioritize the ethical principles* It is important to decide which principle should have highest priority in a particular case. Is respect for persons the most important consideration? Is the usefulness of the data to public health more important? For example, in a benign survey, does respect for persons need to have top priority? In a particular research project, is there a moral imperative in this case and thus no room for compromise? The ethical principles one considers in making a decision can be viewed as *prima facie* rules or obligations in that they have a broad justification and intuitive appeal. However, since it is often not possible to follow all of these principles at the same time, one must decide which one takes precedence and why. For example, a researcher receives a call from a participant in his research protocol, who is experiencing what appears to be a serious side-effect. However, he has promised to attend a university committee meeting. Although he has *prima facie* obligations to keep his promises and to promote patient/subject safety, his duty to his patient overrides his duty to keep his promise.
6. *Consider alternative rankings and alternative solutions* Poor decision making often occurs as a result of bias and ignorance. In order to avoid this problem it is important to be open-minded and to consider different points of view.

7. *Make a well-reasoned decision* Whenever one deals with public health or research issues, one often needs to be able to articulate and communicate decisions to clients or patients, employers, government agencies, and, if necessary, the public. Carrying out a process of reasoning helps one to express and justify ethical decisions and actions. Moreover, one needs to articulate and communicate the decision to those affected by it, which must also be defensible in public if necessary. This is usually an important decision point for choosing right or wrong.

ETHICAL ANALYSIS

We have described an ethical decision-making procedure and will proceed to show by example how to analyze two cases. The method we will follow in our analysis is not unique and one need not conform to it to reach an ethical decision. The method may also appear mechanical in nature because of its simplicity. Nevertheless, we believe this method models the process of ethical decision making.

Example 1

A researcher is at a social gathering in his local community and, after a few cocktails, tells a neighbor about one of the research subjects in his study. The researcher gives sufficient identifying characteristics including the disease diagnosis to enable the neighbor to recognize the person.

Analysis

Step 1 We assume that we have all of the necessary information for this case.

Step 2 His options were to: (1) say nothing about his research; (2) say something about his research but provide no information that would identify subjects; or (3) disclose information that could identify participants.

Step 3 The parties (stakeholders) in order of their importance for this care are
- a. Patient (primary)
- b. Researcher (primary)
- c. Neighbor (secondary)

d. Researcher's institution (secondary)
e. The profession (tertiary)
f. Society (quaternary)

Step 4 Options 1 and 2 do not violate confidentiality, but option 3 does. Confidentiality takes precedence in this case and is supported by the principles of autonomy and nonmaleficence. Breaking confidentiality violates the subject's autonomy by taking away his or her control over personal information and it harms the patient (the primary harm), the institution (secondary harm), the profession (tertiary harm), and society (quaternary harm).

Step 5 The ethical principles' priority in this case is as follows:
a. Autonomy
b. Nonmaleficence

Step 6 There are no acceptable alternative rankings for this case.

Step 7 The researcher has violated the confidentiality of his research subject. It is an unethical act and, in this case, he may have broken some state's confidentiality of patients' records laws. The researcher has harmed the patient by revealing confidential information to a friend, which may cause psychological harm, stigma, job discrimination, or ostracism. The researcher's act may also harm his institution by potentially tainting it with having less than the high professional standards it has set. The act may also harm the researcher's profession by not following professional standards. Finally, the act, if multiplied, could have a negative impact on the mores of society.

We intentionally gave this simple example to illustrate the ethical implication of the act. Other examples may be more complicated and less clear in identifying the ethical terrain.

Example 2

A 40-year-old unemployed cancer patient goes to the emergency room because of serious complications from stomach cancer. He is referred to the cancer specialist at the hospital for further evaluation. The specialist, after careful examination, talks to the patient about a new drug he is testing for a certain company, which could help him. The specialist tells him that this treatment will not cost him anything. He says that, if the patient wants to consider enrolling in the protocol, he will send him to the nurse coordinator of the project and he would have to sign papers. The research protocol for this drug had been approved by the IRB. The patient visits the nurse coordinator, who explains in detail the informed consent process and content of the form, and about all of the alternative therapies, including how long they would last and how much they would cost. The patient tells the nurse that he is poor. She dwells on how these other treatments are expensive and that the new one is free. The patient enrolls in the study. The consent form does not mention the fact that this drug has been shown to have slightly higher toxicity in animals.

Analysis

Step 1 We assume that we have all of the necessary information for this case.

Step 2 The options involve the different items of information that the nurse and researcher/physician communicate to the patient and how they communicate them. Some of these include (a) tell the patient about the toxicity in animals or (b) don't tell the patient; (a) emphasize the high costs of alternative treatments or (b) do not dwell on those costs.

Step 3 The parties (stakeholders) in order of their importance for this case are

a. Patient (primary)
b. Researcher/physician (primary)
c. Nurse coordinator (primary)
d. Researcher's institution (secondary)
e. Researcher's profession (tertiary)
f. Society (quaternary)

Step 4 The issues involved and their placement among the ethical principles are as follows: first of all, to do no harm to the patient; secondly, to benefit the patient by helping him to obtain the best treatment for his condition; thirdly, to promote the patient's autonomy by telling him all of the relevant information and by allowing him to make a free (noncoerced, nonmanipulated) choice; and fourthly, to promote justice by enabling the patient to obtain access to health care.

Step 5 The ethical principles' priority in this case is as follows:

a. Nonmaleficence
b. Beneficence
c. Autonomy
d. Justice

Step 6 One may argue that autonomy takes precedence over nonmaleficence or beneficence in this case. One may also argue that justice should have a higher priority because, if health care were available to this patient, the issue of what information should be delivered to him may become less important.

Step 7 The nurse coordinator (and possibly the physician/researcher) violated two ethical principles. The fact that the nurse dwelt on the cost of the drug knowing full well the inability of the patient to pay for treatment was a method of coercion and thus violated the patient's autonomy. The nurse also violated the "do no harm" principle because toxicity, even though only slightly higher, should have been discussed for the patient's consideration. We placed "do no harm" as the first priority because, if harm to the patient ensues, it will become a high priority because of not telling the patient. One can argue that autonomy should be the first priority in this case and we would have no objection. In their defense, the nurse and the physician may argue that they were simply trying to do what was best for the patient (beneficence), that they were trying to help him to gain access to care (justice), and that the knowledge gained from his participation would benefit others patients and society (justice/utility).

Quiz: Choose the Best Response

1. The Nazi experiments before and during World War II were conducted on:
   a. Mentally retarded children
   b. Mentally ill people
   c. Jews
   d. Disabled people
   e. All of the above

2. The Nuremberg Code was written by the:
   a. Three allies invading Germany: Americans, Russians, and British
   b. Presiding judges in one of the trials
   c. World Medical Association
   d. American Medical Association
   e. None of the above

3. The Nuremberg Code states:
   a. The consent of a human subject is essential
   b. The consent of a human subject is absolutely critical
   c. The consent of a human subject is absolutely essential
   d. Voluntary comprehended consent is absolutely essential
   e. The voluntary consent of a human subject is absolutely essential

4. The Helsinki Declaration was first written in:
   a. 1947
   b. 1964
   c. 1973
   d. 1983
   e. 1996

5. The Tuskegee syphilis study involved the:
   a. Introduction of the syphilis virus into 400 African–American men
   b. Introduction of the syphilis virus into 200 African–American men and women
   c. Observance of the presence of syphilis in 400 African–American men
   d. Observance of the presence of syphilis in 200 African–American men and women
   e. Effect of penicillin on syphilis in 400 African–American men

6. The National Commission was appointed immediately after newspaper reports about:
   a. Willowbrook experiments with mentally retarded children in New York
   b. Radiation experiments on mentally retarded school children in Boston
   c. Dr Beecher's revelations in the *New England Journal of Medicine*
   d. The Tuskegee syphilis studies
   e. All of the above

7. In every case the four ethical principles are prioritized in the following order:
   a. Autonomy, justice, beneficence, and nonmaleficience
   b. Justice, autonomy, beneficence, and nonmaleficience
   c. Beneficence, autonomy, nonmaleficience, and justice
   d. Nonmaleficience, beneficence, justice, and autonomy
   e. None of the above

References

Advisory Committee on Human Radiation Experiments (ACHRE) (1995) Final Report, stock number 061-000-00-848-9. Available from: Superintendent of Documents, US Government Printing Office, Washington, DC. Tel: (202) 512-1800; Fax: (202) 512-2250.

Beecher H (1966) Ethics and clinical research. *New Eng. J. Med.* 274, 1354–1360.

Bean WB (1977) Walter Reed and the ordeal of human experiments. *Bull. Hist. Med.* 51, 75–92.

Caplan AL (1992) When Medicine Went Mad – Bioethics and the Holocaust, pp 1–359. Totowa, New Jersey: Humana Press.

Capron AM (1989) Human experimentation. In: Medical Ethics (RM Veatch ed.), p 127. Boston, Massachusetts: Jones and Bartlett.

Katz J (1972) Experiments with Human Beings. New York: Russell Sage Foundation.

Muller–Hill B (1988) Murderous Science, pp 1–208. Oxford, UK: Oxford University Press.

National Bioethics Advisory Commission (NBAC) (1998) Research Involving Persons with Mental Disorders That May Affect Decision Making Capacity, vol. I: Report and Recommendations. Rockville, Maryland: NBAC. Available at: http://www.bioethics.gov

National Commission for the Protection of Human Subjects of Biomedical and Behavioral Research (1979) Belmont Report: Ethical Principles and Guidelines for the Protection of Human Subjects of Research. Washington, DC: US Department of Health, Education, and Welfare.

President's Commission for the Study of Ethical Problems in Medicine and Biomedical and Behavioral Research (1983) Summing Up. Washington, DC: US Government Printing Office.

Proctor R (1988) Radical Hygiene-Medicine Under the Nazis. Cambridge, Massachusetts: Harvard University Press.

Veatch RM (1989) Medical ethics: an introduction. In: Medical Ethics (RM Veatch ed.), pp 1–26. Boston, Massachusetts: Jones and Bartlett.

World Medical Association (WMA) (2000) Helsinki Declaration: Recommendations Guiding Physicians in Biomedical Research Involving Human Subjects. Helsinki: WMA.

Chapter 3

Current Federal Regulations, the DHHS, and the FDA

The revelation in the 1960s and early 1970s of several serious incidences of placing human subjects at risk led the Government to appoint the National Commission. Soon after, 45 Code of Federal Regulations Part 46 was enacted. The Regulations underwent several modifications; the latest was in 1991 and is now known as the Common Rule. This governs all research supported by the federal government, especially that funded by the Department of Health and Human Services (DHHS). The Common Rule is enforced at the DHHS by the Office for Human Research Protections (OHRP), formerly the Office for Protection from Research Risks (OPRR), through "assurance" agreements signed between the OHRP and the research institutions. Recently, the OHRP has instituted the Federal Wide Assurance (FWA), which replaces all other forms of assurance. The federal regulations require that each institution conducting research with DHHS funds must have IRBs in place to pass judgment and assure compliance with 45 CFR 46, or its Food and Drug Administration (FDA) equivalent, on proposed research protocols using human subjects. IRBs are bound to conduct a continuing review at least once per year and must report certain serious IRB related events to the DHHS.

Learning Objectives

1. Why the Common Rule (45 CFR 46) was enacted
2. What is the OHRP?
3. How do research institutions comply with 45 CFR 46 and 21 CFR 50 and 56?
4. How many types of assurances are available between the OHRP and research institutions?
5. What events should IRBs report to the DHHS?

In the 1960s and early 1970s the media exposed several serious incidences of the violation of human research subjects' rights, such as those of the Tuskegee syphilis studies and those at Willowbrook. After revelation of the Tuskegee syphilis study, as mentioned elsewhere, the Ad Hoc Advisory Panel was formed in 1973. A consequence of the Panel's report was the passage of the National Research Act of 1974. This Act established the National Commission for the Protection of Human Subjects of Biomedical and Behavioral Research, which issued in 1979 what has become known as the Belmont Report (National Commission, 1979). The Act required for the first time the creation of the OPRR and the system of IRBs.

In 1981, for the first time, regulations were promulgated for the protection of human subjects, known as the 45 CFR 46; since then, these rules have been amended five times. In 1991, the final regulations, signed by 16 federal agencies, became known as the Common Rule, which applies only to federally supported research with human subjects (for an excellent presentation of its history, see ACHRE, 1995; OPRR, 1993). To date, there are no rules for privately funded research with human subjects except for those to be filed with the FDA that are associated with drug approval for marketing in the USA and to be discussed later. Privately funded research is covered by the 45 CFR 46 if the institutional "assurance" agreement so states. Critics point out that this is in sharp contrast to the National Animal Welfare Act of 1966, which protects the use of animals in research regardless of the source of funding. The NBAC has recently recommended adoption of: "A unified comprehensive federal policy embodied in a single set of regulations and guidance" for all research involving human subjects (NBAC, 2000).

The Common Rule provides for the establishment of IRBs in research institutions to review and monitor compliance with the federal regulations (OPRR, 1993). The FDA has its own regulations, codified in 21 CFR 50, which deals with the protection of human subjects and with informed consent, and in 21 CFR 56, which deals with IRBs. The FDA is mostly involved in the drug approval process and data submitted in support of investigational new drug (IND) applications must comply with 21 CFR 51 and 21 CFR 56. The FDA has additional regulations pertinent to IRBs and human subjects such as parts 312 of the IND application, 812 of Investigational Device Exemption, and 860 of Medical Device Classification procedures.

The OHRP is the federal agency designated by the DHHS on behalf of the Secretary of the DHHS to enforce 45 CFR 46 and is thus responsible for monitoring compliance with 45 CFR 46 by research institutions and their investigators. The OPRR, until late 1999, was a unit of the NIH, located in the Director's Office. This arrangement came under severe criticism because the NIH was the funding agency for nearly all federal dollars for research with human subjects. As of late 1999, the newly created OHRP became a part of the DHHS and reports to an Assistant Secretary of Health and Human Services. Some critics may still claim that the DHHS both funds and monitors compliance, and thus a conflict of interest may still exist. The report of the National Bioethics Advisory Commission (NBAC, 2001) recommends the creation of the National Office of Human Research Oversight. The Common Rule or the DHHS regulations provide, in addition to the establishment of IRBs and the OHRP, several other important mandates. Institutions engaged in research must give assurance of compliance to the OHRP, which has initiated the FWA to replace all other assurances, and broadens its domain to cover all research supported by federal government and all researchers that have signed the Common Rule. The assurance document holds the institution responsible for protecting human subjects enrolled in research. It also defines research as "A systematic investigation (i.e. the gathering and analysis of information) designed to develop or contribute to generalizable knowledge" (45 CFR 46.102d). The definition of research covers even a single patient if the intent is to collect data systematically. There is a distinction between research intent and innovative treatment practices. Independent investigators and private practitioners who are not affiliated with the research institute need to obtain the institution's IRB approval.

DHHS REGULATIONS

All research projects carried out with funds provided by the DHHS must comply with 45 CFR part 46. The DHHS regulations apply to the protection of all human subjects who are enrolled in research that is supported in whole or in part by direct or indirect funding (such as the provision of facilities or drugs) anywhere in the world. Researchers at all foreign sites who are conducting research involving human subjects with DHHS funds must comply at a minimum with 45 CFR 46 or its equivalent. Other well-known codes of practice such as the Nuremberg Code and the

Helsinki Declaration are not acceptable substitutes for DHHS regulations. Even though these codes are acceptable as background information, they are not framed in terms of realistic and practical regulations.

The DHHS recently appointed the National Human Research Protections Advisory Committee to advise the Secretary and the Director of the OHRP on all issues related to human research protection. The Committee is working to address the new and numerous issues involved.

Assurances

The regulations are implemented through written assurances made by the research institutions to the OPRR. These assurances are negotiated by the institution and the OPRR. Originally there were three types of assurances: single project assurance, multiple project assurance, and cooperative project assurance. As we have mentioned earlier, the OHRP has wisely scrapped all three types of assurance and replaced them with FWA. This covers all federal agencies that are signatories to the Common Rule, and is a simplification that is greatly welcomed by research institutions.

When the OHRP determines that an institution has violated its assurance obligations, it can suspend or terminate the assurance. This action may result in the stoppage of all federally funded research in the institution until the OHRP action is lifted. In addition to the authority to suspend and terminate assurance, the OHRP can impose varying degrees of action ranging from additional reviews, remedial steps, additional training and education, the imposition of certain restrictions, and the removal of a specific protocol or institution from FWA.

The regulations provide for the creation of IRBs, which must have a minimum of five members, one of whom must be from the community and not connected with the research institution. However, IRBs usually consist of 15–25 members with one member from the community. They are required to retain the minutes of their deliberations for 3 years. In addition, IRBs must comply with whatever the assurance agreement has stated, such as in some cases to retain the research protocol for the same period of time.

Continuing Reviews

IRBS are required to conduct reviews of research protocols, usually a minimum of one per year, as long as any data are being collected, even if the subjects are no longer involved. In expedited reviews, IRBs can follow the conditions agreed upon for the expedited review process. A continuing review should include the status of the protocol and details of the data collection and related activities.

Exemptions

Exemptions according to 5 CFR 46.101b provided for in the federal law are not applicable for so-called "masked" studies. None of these exemptions can be applied to special populations such as fetuses, pregnant women, human *in vitro* fertilization, and prisoners. If data, documents, records, and related items are collected prior to the start of a research protocol that includes human subjects, they may be exempt from federal regulations.

Reporting to the DHHS

The regulations require that IRBs report the following events to the DHHS (45 CFR 46):

1. Changes in the membership of the IRB
2. Serious or continuing noncompliance with the regulations: IRBs need not report any or all noncompliance with the regulations, but they must report noncompliance when their own written procedure requires such events to be reported to the DHHS
3. Any unanticipated problems involving risks to human subjects or other events related to the project
4. Any suspension or termination of IRB approval for a research project

There is some debate whether IRBs should also report to the DHHS all adverse events reported by principal investigators.

FDA REGULATIONS

In contrast to the DHHS, the FDA does not sponsor research programs, but regulates drugs, biological products, medical devices for human use, and food and color additives for which approval is being sought for marketing in the USA (21 CFR parts 50, 54, 56, 312, 314, and ICH guidelines). The various provisions regulate the following: 21 CFR part 50 titled "Protection of human subjects"; 21 CFR part 54 titled "Financial disclosure by clinical investigators"; 21 CFR part 56 titled "Institutional review boards"; 21 CFR part 312 titled "Investigational new drug applications"; and 21 CFR part 314 titled "Applications for FDA approval to market a new drug or an antibiotic drug." The lack of research sponsorship by the FDA reflects some of the regulatory differences between it and the DHHS. However, the FDA has a stricter mandate to review all applications regarding regulated products to insure their safety and efficacy from the start of the research protocol to the marketing stage of the product. The FDA also has a great deal of latitude in cooperating with sponsors as well as placing severe criminal sanctions on investigators and sponsors for violating federal regulations.

The FDA's jurisdiction in regulating clinical trials is a result of the requirement for IND applications. It is necessary for a sponsor to submit an IND application to the FDA in order to receive approval to start testing the drug on human subjects. The submission is a detailed application to the FDA that must conform to all of the FDA regulatory requirements. New drug investigations have three phases. Phase I concerns the first introduction of a drug into humans. These studies are very closely monitored and are usually conducted in inpatient settings on healthy volunteers or patients. The purpose of Phase I is to determine the pharmacokinetics of the drug and its potential toxicity with increasing dosage. Phase I clinical trials are usually conducted on 20–80 subjects. Phase II includes several hundred patient subjects. These trials are closely controlled and monitored for short-term side-effects. Phase III studies involve several hundred to several thousand patients. Their purpose is to determine the safety and effectiveness of the drug and the overall risk–benefit relationship.

Some of the key characteristics of FDA regulation are

1. The FDA allows modifications or waivers of the informed consent procedure only for the emergency use of a test article. There are, however, specific requirements concerning how and when an emergency designation can be made for a test article.
2. Some IND applications are exempt from FDA regulations. However, these exemptions do not absolve the protocol from the requirement of an IRB approval.
3. Each investigator or subinvestigator (e.g. residents and associates) is required to sign Form FDA 1572 as part of an

IND approval. This obligates the investigator, under the threat of criminal penalties, to conduct the study properly.

4. FDA regulations allow for "single patient use" outside a clinical trial—when no other treatment is available.
5. FDA regulations mandate a sponsor's responsibilities even though the clinical trials are conducted outside the sponsor's facilities and jurisdiction.

   Regulation #5 refers to when the sponsor outsources clinical trials to clinical research organizations (CROs) or site research organizations (SROs). Some of the sponsor's responsibilities are in insuring the qualifications of the investigator, monitoring the trial's conduct, insuring appropriate regulatory filing of the application with the FDA, and quality assurance of the tested article during clinical trials, such as shipment, disposition, and compliance with good clinical practices and good laboratory practices for prior animal toxicological data. The sponsor may transfer, in writing, all or part of these responsibilities to another organization such as a CRO (21 CFR 312.52). To be valid, the transfer of each responsibility shall be described specifically in the transfer agreement. In addition, the sponsor "shall monitor the progress of all clinical investigations being conducted under its IND" (21 CFR 312.56). Moreover, when the sponsor discovers that an investigator is not in compliance with the IND signed agreement, the sponsor shall discontinue shipment of the investigational drug. Furthermore, the sponsor must monitor the evidence of safety of the new drug and report this information to the FDA, as required by 21 CFR 312.32. When the drug presents a significant risk to research subjects, the sponsor shall discontinue the study and notify the FDA, IRBs, and other investigators.
6. The FDA requires that investigators should report all adverse events (AEs) from drugs and biological product research to sponsors. Investigators are also obligated to report to the IRB any injuries, deaths, and unanticipated serious AEs.
7. The FDA requires financial disclosure. FDA regulation reviews compensation that could affect the outcome of the study. Investigators (and their immediate family members) shall report to the FDA whether their financial interests trigger the reporting requirement.

8. The FDA, by means of issuing a guideline, adopted the International Council on Harmonization (ICH) *Guidelines for Good Clinical Practice and Clinical Safety Data Management.* The FDA guidelines have four categories of AEs.
   a. AE (or adverse experience): any unfavorable and unintended sign related to the drug tested
   b. Adverse drug reaction (ADR): all noxious and unintended responses to the drug
   c. Unexpected ADR: an adverse reaction not consistent with the product information
   d. Serious AE or ADR: an adverse event that may significantly alter drug development; a serious adverse event can be any of the following
   i. Death
   ii. Life-threatening event
   iii. Inpatient hospitalization or prolongation of existing hospitalization
   iv. Persistent or significant disability/incapacity
   v. A congenital anomaly/birth defect

The FDA reporting time-frame requires that any fatal or unexpected life-threatening ADRs be reported by the sponsor to the FDA as soon as possible but no later than 7 calendar days from the time of knowledge. A full report should be forthcoming within 8 calendar days. All other serious, unexpected ADRs that are not fatal or life-threatening must be reported by the sponsor as soon as possible but no later than 15 calendar days.

Both the DHHS and the FDA require the reporting of serious AEs in the categories of death, a need for several days' hospitalization, and significant disability.

DATA SAFETY MONITORING BOARD

Since 1979 the NIH had suggested that clinical trials should have some form of data and safety monitoring process. In 1998, the NIH announced the policy that every multisite clinical trial should have a data safety monitoring board (DSMD). DSMBs are oversight bodies that monitor on a regular basis the safety of Phase III clinical trials. In 2000, the NIH added the need for DSMBs for Phase I and II clinical trials. Investigators must submit monitoring plans that vary according to the potential risks, the complexity, and nature of the trial. The monitoring procedure consists of detailed plans for reporting adverse events to the IRB, the FDA, and the NIH. It is preferred that DSMB members are not associated with the trial and are free of conflict of interest. They should have expertise in the field of the particular trial concerned, in addition to expertise in bioethics and biostatistics.

To our knowledge, the FDA has no specific requirement for the setting up of DSMBs. However, sponsors may choose to put one in place, especially if they are trying to manage risk. In pretrial discussions with the FDA (via a Consumer Safety Officer), the sponsor or the FDA may well consider utilizing a DSMB as a proactive/precautionary measure. Clearly, the phase of the trial, the degree of novelty of the test article, the experience in preclinical work, and doubtless any public policy, legal or ethical concerns, would dictate a decision on the use of such a procedure. There have been discussions that all clinical trials should be required to have a DSMB. The next few years will see several changes in the overall oversight and safety issues in clinical trials.

Quiz: Choose the Best Response

1. The Common Rule is related to the:
   a. National Research Act
   b. 45 CFR 46
   c. Overwhelming majority of federal agencies
   d. Fact that it does not apply to privately funded research
   e. All of the above

2. The Common Rule provides for the:
   a. Establishment of the National Commission
   b. Creation of an IRB in each institution that uses federal funds for research using human subjects
   c. IRBs to have jurisdiction on all research protocols in research institutions
   d. FDA to abide by it
   e. None of the above

3. At present, the OHRP reports directly to the:
   a. Director of the NIH
   b. The FDA Commissioner
   c. The Secretary of the DHHS
   d. The Assistant Secretary of the DHHS
   e. The Deputy Secretary of the DHHS

4. The OHRP uses primarily the following method to promulgate compliance with 45 CFR 46:
   a. Random unannounced site visits
   b. A negotiated agreement between the OPRR and the institution of how the institution will comply with 45 CFR 46, called "assurance"
   c. The establishment of IRBs
   d. None of the above
   e. All of the above

5. Research funded by the federal government:
   a. Is funded by the DHHS
   b. Is systematic
   c. Contributes to generalizable knowledge
   d. Is conducted on one or more patients
   e. All of the above

6. Research projects involving human subjects that are funded by the DHHS must comply with 45 CFR 46 if they are:
   a. Used in a drug application
   b. Only if carried out within the USA
   c. Exempt if they use less than 50% DHHS funding
   d. Exempt if they use only drugs or facilities
   e. None of the above

7. The composition of IRBs should consist of:
   a. A minimum of five members, all appointed by the institution
   b. A minimum of five members, four of whom are appointed by the institution
   c. A maximum of 25 members, with one from the community
   d. Any number of members as long as one of them is from the community
   e. a and d only are correct

8. IRBs are required by 45 CFR 46 to report to the DHHS the following, *except:*
   a. Changes in membership
   b. Unanticipated risks to human subjects
   c. Suspension of IRB approval for a research project
   d. Noncompliance with their own written procedures that state this must be reported to the DHHS
   e. Noncompliance with the regulations

References

Advisory Committee on Human Radiation Experiment (ACHRE) (1995) Final Report, stock number 061-000-00-848-8. Available from: Superintendent of Documents, US Government Printing Office, Washington, DC. Tel: (202) 512-1800; Fax: (202) 512-2250.

21 Code of Federal Regulations parts 50, 54, 56, 312, 314, and ICH guidelines.

45 Code of Federal Regulations part 46.

International Conference on Harmonisation (ICH) (1998) Guidelines for Good Clinical Practice, adopted by the FDA.

International Conference on Harmonisation (ICH) (1995) Guidelines for Clinical Safety Data Management: Definitions and Standards for Expediated Reporting. Adopted by the FDA: Federal Register.

National Bioethics Advisory Commission (NBAC) (2001) Ethical and Policy Issues in Research Involving Human Participants: Volume 1, Report and Recommendations of the National Bioethics Advisory Commission, Maryland, August 2001.

National Commission for the Protection of Human Subjects of Biomedical and Behavioral Research (1979) Belmont Report: Ethical Principles and Guidelines for the Protection of Human Subjects of Research. Washington, DC: US Department of Health, Education, and Welfare.

Office for Protection from Research Risk (OPRR) (1993) Protecting Human Subjects: Institutional Review Board Guidebook. Washington, DC: US Government Printing Office.

Chapter 4

Informed Consent

Informed consent, Principle I of the Nuremberg Code, centers on the protection of the autonomy of individuals, including protection from harm. This requires that the process of informed consent must insure comprehension, voluntariness, and disclosure in order to maximize protection from harm.

In terms of twentieth-century bioethics, informed consent issues are discussed with reference to "principles," which are then translated in an applied manner by IRBs and implemented by those actually conducting the research and, in turn, in contact with the human subjects.

In order to assure proper informed consent, there are certain critical elements, including the purpose of the study and how it relates to the potential subject's underlying condition and the impact on his or her well-being; what will be asked of that person in addition to what he or she will experience (the procedures of the study); what risks and benefits the person can expect; and what alternative treatments are available to the potential subject apart from in a research setting.

Core informed consent issues range from compensation for participation and the recruitment of subjects to the confidentiality and privacy of individuals' data. Most recently, a range of informed consent-based concerns have been raised in both public and private forums, including issues emerging in relation to the frontiers of scientific research. Cases involving genetic research demonstrate the necessity of applying ethical principles and the challenges that IRBs face in determining what constitutes adequate human research subject protection.

Learning Objectives

1. Informed consent in the context of ethical principles
2. Core informed consent elements and issues illustrated:
   a. Informed consent
   b. Compensation for participation
   c. Genetic research consent

Public awareness of the abuse of human subjects who are participating in research is often promoted as a result of crises, challenges, or major events, whether domestically or internationally.

Logically, the development of applications of ethics is spurred on by these same events, either to assist in resolving the crisis or challenge or to strengthen national and international regulations. From Nuremberg (post-World War II trials of Nazi "scientists") to the ICH (the more than 5-year effort in the 1990s of drafting clinical research guidelines for signatory countries), what has come to be known as bioethics has been defined through "principles." These principles emphasize freedom of individual choice and the appropriate acknowledgment by an individual to participate in clinical research. They are the core of both "informed" and "consent."

Ethical Review Bodies and IRBs (hereafter collectively referred to as IRBs) apply those principles in making determinations related to human subject protection. However, translating ethical principles into working reality is not only a challenge for IRBs, it is also a significant challenge for investigators and study staff at the investigative site. In the USA, as elsewhere where research is being conducted, the goal is to convey the required, elements of informed consent (based in federal regulations and international guidelines) in a fashion that is understandable, having ethical principles as further guidance. To be sure, this requires educated and sensitive staff, not only relative to the "science" of the research being considered but also in regard to the principles surrounding critical issues from the subject's perspective: justice, beneficence, and autonomy. Equally important is the potential subject's trust of the researcher and the staff.

Prior to the administration of the consent process to the prospective subject, drafting and reviewing the informed consent form document is a task shared by all key personnel in the clinical research process. If the sponsor of the research is involved in commercialization of the drug, device or biologic agent being evaluated, then the sponsor (or the sponsor's representative, such as a CRO), the site representatives, and the IRB are all involved in the drafting of the document. Authority for final approval of the document to be used to enroll subjects into a study, however, resides with the IRB. In a research institution (e.g. university), the clinical investigator, with the help of the institution's IRB and/or office of human research protection, assists in the formulation of the informed consent document.

Typically, the sponsor of the study (or CRO) drafts the informed consent document. It is then reviewed by the investigator and submitted to the IRB. IRB members and participants review and deliberate on the appropriateness and adequacy of the information presented in the document. If modifications are made by the IRB, the sponsor and the investigator review the changes. Once there is agreement, the document is formally approved and used by the research personnel authorized to do so at the investigation site. Subsequent amendments to the document may be required by changes in the design of the study (the protocol) or as a result of new information uncovered over time which may impact on issues relating to subject protection. Again, in university settings, the major burden for drafting the informed consent document falls on clinical investigators and/or their staff, such as the study director. IRB staff review the document and offer modifications if needed. The consent document is then submitted to the IRB with the rest of the study protocol application.

An important aspect of the informed consent document is that the research protocol and the informed consent document *and* process must not contradict one another. That is, information in the document and information provided by study staff must be consistent with the study design as defined in the protocol, especially its purpose, the study procedures, and the risks and benefits.

The informed consent document may be viewed as a script from which those administering the consent process (typically study coordinators and investigators) work in order to present the consent form fully and clearly to potential volunteers.

Although the regulations and ethical guidelines suggest essential elements to be included in an informed consent document, the list below outlines the most important points.

- The voluntary nature of the potential subject's participation
- The purpose of the study and how it relates to a potential subject
- What the person will experience (the procedures of the study)
- What risks and benefits the participant can reasonably expect
- Whom to contact for answers relating to the study and the participant's rights
- Circumstances under which the subject's participation may be withdrawn
- Alternative treatments not involving research that are available to the potential subject

- Use of wording that minimizes ambiguity and exculpatory language while maximizing comprehension
- Information relating to who may access the subject's information and the data obtained in the course of the research

Discussion surrounding informed consent typically focuses on these core pieces of information. However, the potential subject must be given the opportunity to consider the document as a whole in its written form. This includes having adequate time to evaluate participation and the opportunity to have any questions answered. Family members, friends, and/or caregivers or legal guardians may also be involved in the consent process, even in the investigator's office. The keys to obtaining appropriate informed consent are in the allocation of adequate time and the availability of knowledgeable personnel to answer research related questions.

It is interesting to note, however, that while the Nuremberg Code ascribes to the principle of obtaining of informed consent as a cardinal rule for the ethical inclusion of human subjects into research, an argument has been made that informed consent may not be necessary or sufficient for the ethical conduct of research (Emanuel et al., 2000).

INFORMED CONSENT ELEMENTS AND ISSUES

Some of the more significant controversies surrounding informed consent have to do with potential volunteers who are not or may not appear to be physically, emotionally or mentally capable of understanding what is being presented. These individuals may range from those suffering from brain impairment to those with psychosocial disorders. More on this issue is found in Chapter 7.

Another area of controversy is that of compensation for those participating in research. In 1999, the NIH conducted research on this topic, which in turn provoked important discussions. Fundamentally, IRBs deliberate on a case-by-case basis to determine whether the proposed payment or inducement is acceptable. They follow the FDA guidelines as found in the updated FDA information sheets (FDA, 1998, p 31).

The basis for their decisions evokes the additional ethical principle of distributive justice (from the Belmont Report (National Commission, 1979 and other documents on research ethics), meaning that those who share the benefits must also share the risks and vice versa); as well as an extension of the principles of autonomy and permission, the avoidance of unacceptable inducements for subject participation (e.g. absence of coercion).

In a wider context, during the 151st Council Session in Ottawa (2000), Canada, the Medical Ethics Committee of the World Medical Association proposed revisions to the Helsinki Declaration, in which the following principle, relating to the payment of subjects in clinical research, is to be found as part of Section 22, on "Informed consent":

> There must be no coercion, constraint, duress, unjustified deception or undue influence [of potential subjects]. *Material inducements should be limited to reimbursement for out-of-pocket expenses and normal levels of compensation for time, inconvenience or discomfort* (authors' emphasis).

Because access to health care itself is an inducement to participate, especially in developing countries, economically disadvantaged communities, and in circumstances of disease states for which there are no, or not enough, viable alternatives, IRBs must deliberate not only the matter of payment to subjects but also *inducements* to participate. Issues of payment include not only financial gain; they can include modes of recruitment to entice volunteers.

An IRB therefore needs to consider a number of factors in its deliberations about whether a proposed payment for a clinical trial subject is reasonable. These may vary depending on the project being considered; not all are applicable in every instance.

FACTORS IN OBTAINING INFORMED CONSENT

The standard for all investigators and their staffs should be to obtain truly informed consent from potential subjects. In addition to the impact of adequate time and expertise to answer questions, as noted above, other factors may influence the informed consent process, either positively or negatively.

1. Purpose of payment to subjects for participation in research; is it:
   a. To acknowledge the altruistic participation of the subject in the form of an honorarium (e.g. as a thank-you for participating)?
   b. To reimburse for travel expenses incurred during the clinical trial (e.g. for study visits)?

c. To compensate for time and or lost work/wages?
d. To compensate for actions the subject takes or procedures the subject undergoes during the clinical trial (e.g. answering a questionnaire, keeping a diary, giving a blood sample, undergoing radiography or other diagnostic procedures, such as endoscopy)?

2. Level of risk(s) of participating in the study compared with benefit(s); is the benefit or risk as presented to the potential subject:
a. Intended to change the subject's decision-making behavior, but only to the extent of participation in the research; and is it without risk of substantial harm (e.g. answering a questionnaire, keeping a diary, giving a small blood sample)?
b. For the subject to take unwelcome risks of harm (e.g. repeated blood drawing or other invasive procedures without significant long-term risk of harm or mortality, or Phase I studies)?
3. Autonomy of the subject:
a. Is the subject's age material to the financial or other study-related inducement?
b. Does the subject have a dependent relationship with the investigator, study staff, or sponsor (e.g. employee, student, institutionalized patient)?
c. What is the relative financial status of the populations being studied (e.g. for needy persons, a small sum, either relating to the procedure or direct compensation, may have higher leverage on behavior)?
d. If the compensation being provided is either monetary or access to health care, is the subject one whose interests are represented by others who will actually receive the money (e.g. pediatric study in which parents receive payment as well as the child; or the study subjects suffer from Alzheimer's disease and caregivers as well as subjects receive payment)?
4. Nature of the payment:
a. Is it cash, cash-equivalent, or an object (e.g. a toy for a child)?
b. When does payment occur: lump sum or pro-rated throughout the study?
c. Is it excessively weighted to completion of the study or as a bonus for completion? If so, is this an issue of coercion or excessive inducement?

d. Are there special privacy and/or confidentiality issues related to the type or amount of compensation (e.g. the necessity of reporting earnings to the IRS or other federal, state, local authorities)?

5. Equitability:

   a. What is the range of variation of compensation among sites in a multisite study; how wide is the range; what accounts for the range; is there a potential bias for/against the inclusion of certain subjects?

   b. How do community attitudes and local conditions impact the types and amounts of compensation offered in one area versus another?

6. Phase of the study:

   a. What is the relationship between the compensation and the phase of the trial? Is more being offered for a greater level of risk?

   b. Are there particular concerns relating to "professional patients" (especially in Phase I and highly compensated studies)?

Considering inducements as a more encompassing issue than payments, the following must also be taken into account:

- The presence/absence of inducing language, graphics, or other media in a study's or institution's recruitment material
- Appreciation of the community and/or patient population being targeted for participation in the study with respect to subtle as well as overt persuasiveness of an excessive nature, for example, the economics of the community, or the 'desperation' of the patient population.

Many IRBs are concerned about the issue of inducements as a special matter for some communities and patient populations. While they have tended to raise these possibilities in the course of deliberation, they have been hesitant to implement stern restrictions, choosing instead to advise researchers, sponsors, CROs, and/or investigative institutions of their concern in order to increase levels of awareness throughout the clinical research enterprise. Some have argued that payments, especially to healthy volunteers, should be large enough to acquire the needed number of participants. In other words, why should a healthy subject take a risk to benefit corporations? They state that high risk professions such as firefighters and policemen do receive higher wages than their counterparts.

Some IRBs generally do view some types of payment as unacceptable. These have tended to fall into the following categories.

- Payment to secure subject participation when the risks are disproportionate to the benefits (e.g. giving money simply for patient consent)
- Amount of payment believed to be vastly in excess of that required for the necessary time commitment and/or the procedures involved in participating in the study
- Large payments to parents/guardians for a child's or adolescent's participation

One more serious area of concern focuses on issues of privacy and confidentiality. These issues are manifest in matters ranging from recruitment via the use of "call centers" and Web/Internet-based approaches to participation in genetic research.

To conclude our discussion of informed consent, a consideration of genetic research is appropriate. This provides a context not only for examining in practical terms what privacy and confidentiality issues actually exist, but also for what constitutes a research subject being informed. Accordingly, sponsors, institutions, and IRBs each need to pay attention to the following factors.

- Diagnostic and/or screening tests involved, and what the differences are, including limitations of the test/screen results as a predictor of clinical risk as well as risk thresholds and feasibility of population screening
- Results being or not being provided to the subject and the opportunity not to receive test results. If results are provided, what means exist for assessing the psychosocial impact.
- Disclosure, "up-front," if no immediately useful or interpretable information of relevance to a subject is likely to result
- Incidental findings: how these are communicated, if at all, to the subject; assessment of unknown risks; study's procedures to identify risks and inform the subject in the future
- Benefits: study results being informative or not to the individual, to the individual's family, the individual's community; commercial value to the individual
- Risks: impacts on privacy and potentials for misuse of the data (such as personal embarrassment, social stigma,

unintended discrimination, job loss, insurance or promotion denial); stereotyping or stigmatizing of the family or group of which the subject is a part; unknown risks; mental health risks

- Psychological and medical impacts: study's procedures to handle potentially negative impact, including availability or not of genetic and psychological counseling
- Withdrawal from study of self and samples: study's procedures that are in place
- Waiver of rights or control of samples as criterion to participate (i.e. issue of coercion)
- Disclosure of any commercial interests in the research
- Time-frames relating to how long the data will be maintained as well as long-term follow-up requirements
- Secondary use: if samples are retained, issues of access (who has), use (for what purposes), and communication obligations (regarding re-contact of subjects)
- Costs implications: in addition to those within the study, any as a result of the outcomes of the study
- Confidentiality and privacy: procedures that are in place and that there are limits to confidentiality and privacy; identification of data with a specific subject and what restrictions, or not, there are on naming individuals in publications or databanks

In summary, more than one historian has begun to refer to the twentieth century as the golden age of scientific and technological advancement. Whether true or not, in the twenty-first century both public and researchers face the ethical consequences of what has been wrought in the previous century. The public must come to terms with the acceptability of trade-offs of innovative health care and disease state management against the risks of achieving such advances.

Researchers are faced with creating potential "breakthroughs" that must await the development of appropriate protocol designs containing acceptable human (and animal) subject risk. The scientific and clinical research enterprise as a whole is faced with the ultimate innovative challenge: developing technologically-assisted, or simply highly creative, methods for testing new chemical entities and genetic material, as well as "alternative or complementary medicines" in order to improve the predictability of usefulness and limit the use of humans or animals in research.

The key points of this chapter are

- Ethical principles are not only the basis for informed consent, they also direct the process.
- Regulations, deriving in part from ethical principles, require certain elements to be contained in informed consent documents.
- Special issues of informed consent include: the process of fully informing potential research subjects; the matter of compensation for participation; and the contemporary issue of genetic testing as it relates to consent (including ethical concerns about privacy, confidentiality and coercion).

Quiz: Choose the Best Response

1. Informed consent is a:
   a. Document
   b. Process
   c. Regulatory requirement
   d. a and c only
   e. a, b, and c

2. ICH refers to:
   a. US federal guidelines
   b. US federal regulations
   c. Worldwide standards for conducting clinical research
   d. Guidelines for conducting clinical research applying only to signatory countries
   e. None of the above

3. IRBs must deliberate:
   a. Financial payments to subjects
   b. Inducements to participate
   c. Methods for recruiting subjects
   d. All of the above.
   e. None of the above

4. Which of the following is outside the purview of the IRB:
   a. Potential embarrassment to a subject
   b. Commercial value of genetic material
   c. Privacy of patient records
   d. Community attitudes toward clinical research
   e. All of the above

5. The primary constituents of the clinical research process are:
   a. The sponsor
   b. The investigator
   c. The FDA or the NIH
   d. All of the above
   e. a and c only

Case Studies

Case 1

A patient in a physician's office is approached to participate in a clinical research project. The physician also conducts such studies and is doing so as part of this particular project. The physician's study coordinator, an experienced individual, representing the physician/investigator, has reviewed the patient's medical records and has determined, with the physician's oversight, that the patient may qualify to participate. In approaching the patient, the coordinator suggests that the research project involves a new therapy that shows great promise, encouraging the patient to consider participation.

What are the ethical implications in how the study coordinator is beginning the consent process? Is the coordinator acting appropriately so far?

Case 2

In the course of a routine physical examination, a physician discovers an incipient medical condition and asks if the patient would like to participate in a clinical research project involving this condition. The study is being conducted by a close colleague of the physician. As part of this discussion, the physician indicates that participants will receive more detailed medical attention in the course of the study, which will be from a specialist in the field. In addition, there will be compensation to subjects who are enrolled to cover wages forgone for missing work. What the physician does not disclose is that for referring patients who are enrolled into the study, the physician will receive a bonus of $2500.

Under what circumstances may the physician have a responsibility to disclose the bonus to the patient?

Case 3

An announcement indicating that volunteers are needed to test a new toothpaste that whitens teeth appears in a local community newspaper as well as on flyers posted throughout the community, which comprises people with low incomes. The advertisement's headline reads, in large and bold print: "White Teeth for the Rest of Your Life." It also states: "You will receive $100 and free transportation to and from the clinic."

As an IRB participant (member or alternate), what are your concerns? What wording should be used in the informed consent form to indicate compensation?

Acknowledgements

In formulating the summary relating to genetic research, the authors are appreciative of Kathleen Cranley Glass, et al. (1997) Structuring the review of human genetics protocols, Part II: Diagnostic and screening studies. *IRB* 19(3–4), 2–3; and Ada Sue Selwitz (1996) Issues to be addressed in obtaining informed consent involving DNA banking and genetic research. *ARENA*.

References

Advisory Committee on Human Radiation Experiments (ACHRE) (1995) Final Report, stock number 061-000-00-848-9. Available from: Superintendent of Documents, US Government Printing Office, Washington, DC. Tel: (202) 512-1800; Fax: (202) 512-2250.

Beauchamp T & Childress J (1994) Principles of Biomedical Ethics, 4th edn. New York: Oxford University Press.

Capron AM (1989) Human Experimentation. In: Medical Ethics (RM Veatch ed.), pp 125–172. Boston, Massachusetts: Jones and Bartlett.

21 Code of Federal Regulations parts 50, 56

45 Code of Federal Regulations part 46

Council for the International Organizations of Medical Sciences (1993) International Ethical Guidelines for Biomedical Research Involving Human Subjects. Geneva, Switzerland: CIOMS.

DeRenzo EG (2000) Coercion in the recruitment and retention of human research subjects, pharmaceutical industry payments to physician-investigators, and the moral courage of the IRB. *IRB Rev. Hum. Subjects Res.* 22(2), 1–5.

Dunn C & Chadwick G (1999) Protecting Study Volunteers in Research: a Manual for Investigative Sites. Boston, Massachusetts: CenterWatch.

Emanuel EJ Wendler D & Grady C (2000) What makes clinical research ethical? *JAMA* 283, 2701–2711.

Engelhardt HT (1996) The Foundations of Bioethics, 2nd edn. New York: Oxford University Press.

Epstein KC & Sloat B (1996) Drug trials: do people know the truth about experiments? In the name of healing. *The Plain Dealer* Dec 15.

Epstein KC & Sloat B (1996) Drug trials: do people know the truth about experiments? Foreign tests don't meet US criteria. *The Plain Dealer* Dec 17.

Food and Drug Administration (FDA) (1998) Information sheets: Guidance for Institutional Review Boards and Clinical Investigators. Rockville, Maryland: FDA.

Ginsberg D (1999) The Investigator's Guide to Clinical Research, 2nd edn. Boston, Massachusetts: CenterWatch.

Hartnett T ed. (2000) The Complete Guide to Informed Consent in Clinical Trials. Springfield, Virginia: PharmSource Information Services.

Levine R (1986) Ethics and Regulation of Clinical Research, 2nd edn. New Haven, Connecticut: Yale University Press.

National Commission for the Protection of Human Subjects of Biomedical and Behavioral Research (1979) Belmont Report: Ethical Principles and Guidelines for the Protection of Human Subjects of Research. Washington, DC: US Department of Health, Education, and Welfare.

Office of the Inspector General (2000) Recruiting Human Subjects: Pressures in Industry-Sponsored Clinical Research. Available at: http://www.dhhs.gov/progorg/oei

Office of the Inspector General (2000) Recruiting Human Subjects: Sample Guidelines for Practice. Available at: http://www.dhhs.gov/progorg/oei

Sloat B & Epstein KC (1996) Drug trials: do people know the truth about experiments? Using our kids as guinea pigs. *The Plain Dealer* Dec 16.

Sloat B & Epstein KC (1996) Drug trials: do people know the truth about experiments? Overseers operate in the dark. *The Plain Dealer* Dec 18.

(1996) The Nuremberg Code. *JAMA* 276, 1691.

Weiss R & Nelson D (2000) US halts cancer tests in Oklahoma. *The Washington Post* July 11.

Weiss R & Nelson D (2000) FDA faults Penn animal tests that led to fatal human trial; genetic research killed teenager. The Washington Post July 12.

World Medical Association (WMA) (2000) Helsinki Declaration: Recommendations Guiding Physicians in Biomedical Research Involving Human Subjects. Helsinki: WMA.

Chaper 5

Institutional Review Boards

IRBs have historically been associated with large, academic, teaching medical centers where research is a part of the core mission of the institution. The traditional role of the IRB has been associated with the "ethical oversight" of research involving human subjects. While one may speculate intuitively that an IRB is a critical component of any research enterprise, oversight by this committee, whose existence is mandated by regulations, varies widely.

Academic center-based IRBs have been, and continue to be, a staple component of clinical and biomedical research. Because of the rapid expansion of clinical research beyond the universities, especially in the area of drug, device and biologic agent development, independent IRBs have also become established, particularly since the early 1970s.

Overall, IRBs remain an important part of the clinical research landscape, both domestically and internationally.

Learning Objectives

1. The IRB's role(s)
2. How the IRB fits into clinical research enterprise
3. What constitutes an IRB
4. How an IRB operates

The way in which IRBs function has evolved as an extension of the peer review system for reviewing and judging research protocols, specifically with regard to the protection of the rights and welfare of human participants. The pressure to provide greater protection for human subjects increased dramatically after World War II in the light of revelations about Nazi and Japanese human experimentation. Subsequent revelations relating to the mistreatment of elderly people, prisoners, and, most dramatically, African–Americans in the USA as well as cultural unrest in the late 1960s and 1970s, fueled public opinion for legislation.

In 1974, the then Department of Health, Education, and Welfare issued the federal regulations known as 45 CFR 46, for the establishment of IRBs. Subsequent regulations in 1981 provided more detail on the function of IRBs (Levine, 1986) and their role in both privately (pharmaceutical company) or publicly (federally) funded/sponsored research.

This chapter will discuss the structure and function of IRBs, the "crossroads" they now face in the light of emerging issues, as well as the impact on sponsors and investigators; and how investigators and IRBs, as well as their staff, interact in the course of the review process as a study moves toward approval, implementation, continuing review, and final review. For a detailed description of the evolution of the regulations leading to the establishment IRBs, the reader is directed to Chapter 3 or other texts (Levine, 1986) and resources (OPRR videotape series).

THE SETTING

Since their creation, IRBs have been charged with the regulatory mandate to provide oversight for the protection of the "rights and welfare" of human subjects participating in research (see 21 CFR 56, 45 CFR 46 and the Common Rule). The regulations promulgated to empower IRBs use as their foundation the principles articulated in the National Commission for the Protection of Human Subjects of Biomedical and Behavioral Research's report, (the Belmont Report) (National Commission, 1979). However, how IRBs discharge their duties is not clearly spelled out in the regulations, neither is there specific algorithmic, guidance at the present time. As such, there is much variation regarding functional and operational processes among IRBs. The regulations specifically indicate, however, that an IRB must approve research involving human subjects before it may proceed. The regulations also mandate that an IRB must provide ongoing and continuing review of approved research and that the interval for such continuing review shall not exceed a period of 12 months.

Since the late 1990s, the role of IRBs has come under both scrutiny and criticism.

- The 1998 Office of the Inspector General's Report questioned whether current human subject protection practices constitute "a system in jeopardy." A further report less than 2 years later indicated the inadequacy of response to the previous report.
- Federal and state legislation has emerged on fronts ranging from inclusion and privacy to reimbursement, access to information, and ownership rights, offering the prospect of recharting dramatically the course of the relationship between the patient and the clinical research enterprise.
- The inevitable collision of patient advocacy and the brave new world of biotechnology have become staple constituents of social and public policy debates.

THE CURRENT SITUATION

Elsewhere we have suggested that IRBs are one of a society's "mediating structures." [1]The implications for what roles IRBs play are profound, as we will discuss in this chapter.

Clinical research itself is an institution of society. In addition to the promise of finding cures for and further understanding of diseases, research activities offer to society a professional, economic, and social benefit. Clinical trials are an important source of medical care, with a currently large capability, and even larger potential, for community and economic development. Clinical research promises to continue as a significant source of jobs as well as product and service innovation. Health care and biomedical research continue to experience major transitions, most obviously in redefinition and globalization.

Into this complex and sophisticated environment the prospective volunteer is recruited, sometimes in a personal way (via his or her own physician), sometimes in a more distant manner (through various print and nonprint media, including the Internet) (see Chapter 4). Once recruited, the subject becomes part of an orchestrated effort of treatment and observation in the care of a limited number of individuals with whom she or he becomes acquainted: investigator, study coordinator, nurse, and other personnel. The expectations of potential subjects may well range from being entirely limited to free care and treatment to some hope of successful dealing with a physical condition, such as prevention or alleviation.

From the subject's position, somewhere far behind the scenes lies the provider of the treatment entity (sponsor, funding agency), perhaps another intervening organization that is somehow related to running the study (sponsor's representative, university's specific center or organization), and, appearing most abstract, an entity with the impressive title of an institutional or ethical review board. Although the subject may gain confidence by getting to know the study staff and, perhaps, from name recognition of the sponsoring organization, rarely is there an understanding or a modest sense of credibility added by the presence of this protection reviewing entity. But what if enrolled or recently enrolled subjects:

- Are hospitalized for a reason that may be related to the study?
- Are "coerced" either to participate or to continue participation by the threat (real or perceived) of loss of medical benefits to which they are otherwise entitled?
- Misled regarding possible benefits or other dollar or nondollar compensation for participation in research?
- Do not receive promised payments for their participation?
- Have questions regarding their rights and welfare relating to participation in the research?
- Are denied insurance reimbursement for treatment not clearly related to participation in a trial?

There are hundreds of other scenarios – escalating from poor communication to fraud – that actually occur in the conduct of research and generate concerns or complaints, including by the subject.

To whom should the research subject turn for assistance and help? Given the regulatory charge of the IRB, one would expect that to be the logical choice.

If there are inadequate supporting structures, poor protection of humans who are participating in research can be detrimental not only to those who are being served but also to society as a whole. Like airplane crashes, they may be rare, but they can be lethal... and not exclusively to research subjects.

THE ISSUE OF THE IRB'S ROLE: AN IDENTITY CRISIS ALREADY HAPPENING

An IRB is a committee of human beings charged with deliberating the risks and benefits of research. From a regulatory perspective, an IRB's constitution should consist of a range of individuals to include at least five members (some of whom must possess the necessary scientific expertise to review the proposed research), a representative who is otherwise not affiliated with the institution, a nonscientific member, and a representative of the community such that local attitudes can be factored into the acceptability of the research being reviewed. In addition, an IRB should represent gender, age, and racial diversity, which may be appropriately representative of the community in which the research is being conducted. In many major research institutions the membership of IRBs numbers between 15 and 20, while the nonaffiliate member remains at one. This dilution of community representation has come under criticism from patients' advocacy groups.

To inform their conversations and deliberations, participants attempt to take into account four important perspectives: the regulatory, the ethical, the legal, and the scientific. However, IRBs are neither constituted nor directed to be expert bodies in any one of these, much less all four.

At present, IRBs indisputably play a primary role of gatekeeper, for without IRB approval there is no clinical trial to be conducted. Beyond that, there is industry and regulatory debate as to how powerful the role of an IRB should be. For example, in situations of concern, complaint, or downright disagreement involving research subjects (between subject and principal investigator, or among others), are IRBs:

- "Neutral" bodies assisting in negotiations between research subjects, clinical investigators, university administrators, institutions, sponsors, and other parties (ranging from sponsors' representatives, like CROs, to health care insurance providers)?
- Advocates for subjects' protection?
- Arbitrating channels – among various parties – to each other, with the ultimate authority (some would argue too ultimate) to decide?

Each IRB must grapple with these fundamental questions of identity and define its own appropriate response. At the heart of the IRB's response is the question: "What do we as an IRB mean by subject protection?" Such a query rapidly moves into clinical, emotional, and financial discussions and how, if possible, to separate these aspects. An even more taxing round of discussion will ensue should the IRB choose to take on the thorny issue of how intimately and necessarily these aspects are related.

Federal regulatory agencies give IRBs the broad mandate of human research subject protection. Their working model for interactions among IRBs, institutions, and subjects provides a solid, but only one, dimension for trying to understand what really is the IRB's role. The reporting responsibility and relationships among the various participants in the research process is noteworthy. While the regulatory agencies (FDA, NIH, OHRP) have direct oversight and authority over all those involved, IRBs have the responsibility of protecting research subjects through the institution. Therefore, institutions are obligated to the IRBs and the constraints they impose on them in order to assure adequacy of protection including satisfactory and proper informed consent, and ongoing and comprehensive continuing review. In addition to the regulatory responsibility to IRBs, institutions are also obligated to maintain direct, prompt and open communication with the sponsor of the research as described in Figure 5.1.

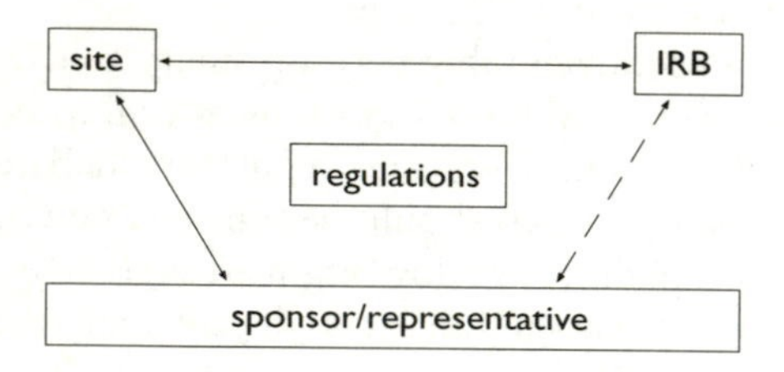

Figure 5.1 Reporting responsibilities

Another dimension is that of the phase and the process stages of a clinical trial. Does the role of IRBs change depending on either the clinical trial phase (I, II, III, or IV) or the clinical trial process stage (pre-initiation, initiation, study conduct, or close-out), or both?

IRBs and their regulators grapple with the extent to which they can and should be constituted to accept and accommodate conflict (serving as an advocate or an arbitration channel) and/or to avoid compromise (serving as a mediator, or being "neutral"). The regulatory mandate is clear: human subject protection, first, foremost, and last. However, to what lengths should IRBs go to protect those in the care of investigators, who are directly responsible to the IRBs?

IRBs are not ombudsman bodies, which are offices that are formally defined and charged to determine and hold specific individuals accountable; or at the very least empowered to resolve disputes with whatever measures are deemed suitable; or are they? Doubtless, most IRBs have members and alternates with individual predispositions towards one or another version of these roles, or an inclination towards situational decision making. Overarching this level of complexity is the topic of variation among IRBs and their individual interpretations of the extent of their mandate. The most common complaint among users (sites, sponsors, sponsors' representatives) of the range of existing IRBs is rooted in their failing to appreciate how a particular IRB sees its own role. However, a significant contributing factor to this madness is self-inflicted. There is insufficient appreciation that, while all the parties conducting clinical research do indeed share a common goal, they do not share a common role.

IRBS' MAKE-UP AND OPERATIONS

An IRB may be made up of as few as five members representing science, the community, and the institution hosting the IRB. Meetings are conducted formally, according to "rules of order." Voting must be formally recorded, together with indications of deliberations, in the minutes of IRB meetings. While "simple majority" rules, regulatory interpretation, as currently imposed by the oversight agencies (the FDA and the OHRP) requires official records of the voting process to reflect the exact numbers voting for, against or separately on each action taken by the IRB.

Many IRBs, especially those at academic medical centers and large community hospitals, have membership in the 20s and 30s or more, making quorums and full attendance difficult to achieve. Added to this may be an inadequate administrative support structure to accommodate the research demands of the institution. The simple procedure of convening a meeting in a timely manner may suddenly become an insurmountable administrative burden.

To understand more about how IRBs function, it is useful to take a bird's eye view of clinical research as it operates at the investigation site level, with specific appreciation that an IRB based in an institution operates very differently from one that is independent of any site affiliation.

Clinical research studies may involve a single physician in one location or up to thousands of physicians across the world. These are known as single site or multisite studies, respectively. There are different types of research sites, including those based in an academic, community health care, or managed care organization setting; a routine clinical practice setting; a clinical research practice setting; or within a multisite organization, often referred to as either trials management organizations or site management organizations. Depending on the institutional setting, an IRB's authority will be academic, hospital, or independently based.

All IRBs require site-specific information (i.e. data on the personnel at the site). This information ranges from research experience to a résumé of the individual(s) who present the study and the consent process to prospective subjects. Of course, IRBs also require information on proposed studies. This includes the study protocol as well as information on the history of and experience with the study article (drug, device, biologic, biotechnologic), as contained, for example, in the Investigator's Drug Brochure, a product insert as approved by the FDA, or other relevant information describing both preclinical and clinical experience. The types of information that are exchanged during the course of a study between IRBs and research sites are illustrated in Figure 5.2.

A picture of an initial IRB review of a study and institutional site is provided by a simplified scenario.

> A presentation by a scientist/physician of the protocol, focusing on the methodology of the study, is followed by a full IRB discussion. Subsequently, a nonscientific member introduces social, community, and ethical considerations as well as issues relating to the consent form and process. These topics, too, are deliberated, followed by a review of the site's credentials. Votes to approve, disapprove, or defer (for further clarification) are taken on the protocol, informed consent form and process, and investigational site.

The role of the staff of an IRB is as critical as that of the IRB itself. After all, the staff are charged with implementing the IRB's decisions and, equally important, gathering and providing information on which those decisions are made. The staff are, by and large, charged with handling the entire process, from taking in information and clarifying it to confirming information and releasing approvals.

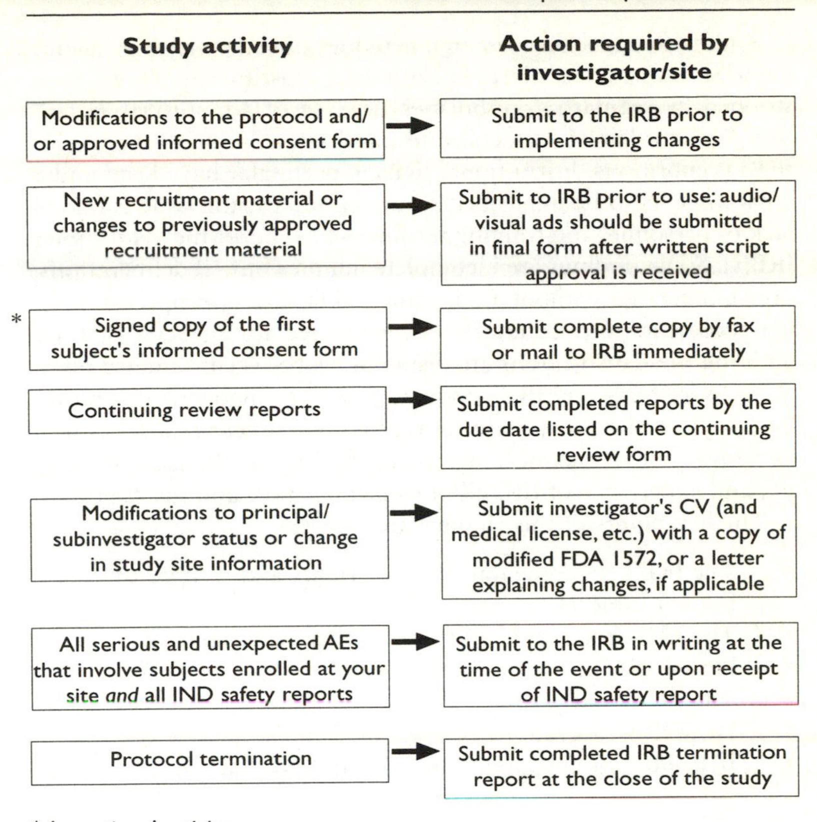

Figure 5.2 IRB contact flowchart

Recent regulatory discussion has reiterated what the consent process entails from FDA, OHRP, and NIH perspectives. It spans recruitment through withdrawal from the study! This means that the IRB must review and approve virtually all forms of recruitment material and campaigns as well as continue to review the institutional site's compliance with regulations, especially the enrollment and retention of research subjects, throughout the duration of the study. In addition, currently, an IRB is also expected to determine changes in the relationship between the benefits and the risks to subjects throughout the course of a study. This means reviewing AEs as defined in the protocol, as well as any new findings submitted in the form of amendments to the protocol by the sponsor or the researcher.

Is it any wonder that research activities of some major academic medical centers and large health care institutions have been stopped by regulatory authorities because of administrative failure? The burden of data collection and information exchange for IRBs is enormous. Institutional IRBs in particular have been under what appears to be intensified scrutiny. In part, the issue is one of lack of personnel and training resources to support the institution's IRB(s). Some findings are incomplete minutes of IRB deliberations, decision making without a quorum, and lack of appropriate member representation. Of equal concern to federal agencies that are auditing these institutions are disjunctions between standard operating procedures and IRB conduct, as well as inadequate continuing review and AE oversight. In recent government reports as well as audits, one of the most significant findings is the lack of training and experience of IRB administrators, staff, and members.

The key points of this chapter are

- The IRB's fundamental charge: "rights and welfare of human subjects"
- The IRB's role: an identity crisis already happening
- How IRBs 'fit' into clinical research
- The IRB's make-up
- How IRBs operate
- IRB shortcomings found in FDA and NIH audits

Quiz: Choose the Best Response

1. The IRB must have at least:
   a. Three members
   b. Four members
   c. Five members
   d. Six members
   e. The regulations do not require a minimum
2. The primary charge of the IRB is to:
   a. Determine the competency of the investigator
   b. Insure that the study coordinator is experienced
   c. Protect human subjects participating in the study
   d. Help the sponsor to meet the stated objectives of study
   e. None of the above
3. In order for the IRB to perform its duties, the Board must:
   a. Determine the competency of the investigator

b. Review the competency of site personnel who are obtaining consent
c. Approve the appropriateness of graphics in advertising
d. Determine acceptable levels of financial inducement
e. All of the above

4. Audits of IRBs by federal agencies have found deficiencies such as:
a. Inadequate minutes
b. Insufficient training
c. Inadequate oversight in multiple areas
d. All of the above
e. None of the above

5. Which of the following is/are *not* an IRB?
a. Independent IRB
b. Hospital ethics committee
c. Promotion and tenure committee
d. More than one of the above
e. None of the above

Case Studies

Case 1
One of the options currently under discussion is to define the role of the IRB by restricting it and turning over responsibility for monitoring AEs and privacy issues, for example, to other types of deliberative bodies (like a DSMB or a privacy board). Create an argument *against* this approach.

Case 2
Now create an argument *for* the approach that multiple bodies, rather than just the IRB, should be involved in human research subject protection issues.

Case 3
Imagine you are setting up an IRB to focus on vaccine trials only. What types of individuals would you have on the IRB to meet the spirit and intent of the regulations? Since the vaccine model of medicine has some severe critics, how would you incorporate their perspective on the IRB? What happens if the scientist members of your IRB vote to approve the protocol and the community members do not?

Notes

[1]In an article by Chesapeake Research Review Inc. in *IRB* (September–December, 1995, pp 12–16) entitled "On being an IRB." According to Peter L. Berger and Richard John Neuhaus, "mediating structures" in society are: "those institutions *standing between* the individual in his private life and the large institutions of public life... Such institutions have a private face, giving the private life a measure of stability, and they have a public face, transferring meaning and value to the megastructures... [of] economic conglomerates... and bureaucracies that administer wide sectors of the society, such as in education and the organized professions (authors' emphasis)." *To Empower People* (DC: American Enterprise Institute, 1977), pp 2–3.

References

21 Code of Federal Regulations parts 50, 54, 56, 312, 314

Dunn C & Chadwick G (1999) Protecting Study Volunteers in Research: a Manual for Investigative Sites. Boston, Massachusetts: CenterWatch.

Levine R (1986) Ethics and Regulation of Clinical Research, 2nd edn. New Haven, Connecticut: Yale University Press.

Office of Inspector General (1998) Institutional Review Boards: a Time for Reform. Washington, DC: Department of Health and Human Services. Also available at: http://www.dhhs.gov/progorg/oei

National Commission for the Protection of Human Subjects of Biomedical and Behavioral Research (1979) Belmont Report: Ethical Principles and Guidelines for the Protection of Human Subjects of Research. Washington, DC: Department of Health, Education and Welfare.

(1996) The Nuremberg Code. *JAMA* 276, 1691.

World Medical Association (WMA) (2000) Helsinki Declaration: Recommendations Guiding Physicians in Biomedical Research Involving Human Subjects. Helsinki: WMA. The latest version is available at: http://www.wma.net/e/policy/17-c_e.html

Chapter 6

Conflict of Interest

Conflict of interest arises when one's private interest is in conflict with the public duty to render an impartial opinion. It is in the nature of human beings to be influenced in this situation. The import of conflict of interest could vary from the benign to the criminal. Therefore, in order to maintain confidence in our institutions, we must design decisional structures that avoid conflict of interest. The use of human subjects in research is an important case in point in which private interests may come into conflict with public interest. Investigators and sponsors can derive a great deal of personal benefit from carrying out research with human subjects. The avoidance of conflict of interest is therefore ultimately in the best interest of society, the investigator, and the sponsor.

Learning Objectives

1. Definition of conflict of interest
2. Prevalence of conflict of interest
3. How are IRBs influenced by conflict of interest?
4. Methods of avoiding conflict of interest

The first-century sage, Hillel, said: "If I am not for myself, who will be for me? If I am only for myself, what am I?" In this statement, Hillel encapsulates the dichotomy of human nature and the struggle between public good and private interests. This dichotomy is part of our nature. These two conflicting interests have been and will remain with human beings as part of our daily lives.

Black's Law Dictionary (Garner, 1999), defines conflict of interest as "A real or seeming incompatibility between one's private interests and one's public or fiduciary duties." With these simple words, conflict of interest is defined as a conflict between private good and public good regardless of the circumstances. It appears in every profession of our society: politics, journalism, science, research, government, and financial institutions. The influence of conflict of interest in all of these professions ranges from the trivial to the criminal act. Management of the conflict between public good versus private interest is important in a democratic society because democracy is built on public trust that is essential to the voluntariness of the populace to be governed. Conflict of interest can have a slow and chronically corrosive effect on democratic values. It can erode the confidence of people in their institutions (Shamoo and Dunigan, 2000).

One may ask, therefore, when does conflict of interest come into play in research? It can be evident in every step of the research protocol, from design, selection of methods, selection of subjects, content of informed consent procedure, IRB submissions, IRB composition and its deliberations and voting, and many more. What concerns us most is the conflict of interest between principal investigators, sponsors, and research institutions, and the outcome of clinical research or drug trials with human subjects. It is not surprising that reports indicate drug efficacy outcomes in clinical trials that are highly in favor of sponsors (Porter, 1992).

PUBLIC HEALTH SERVICE

Since 1995, the Public Health Service (the parent agency of the NIH) expressed concern regarding the influence of conflict of interest on the integrity of research and its potential to erode public trust. Our common sense dictates that conflict of interest, whether for financial gain, career advancement, or promotion and attainment of research funding, can undermine our objectivity. This is especially true if there are direct or indirect financial gains from the outcome of research. Therefore, the Public Health Service requires that institutions should

1. Enforce policy on conflict of interest
2. Inform all investigators
3. Report to the awarding agency the existence and management of conflict of interest

IRBS AND CONFLICT OF INTEREST

Currently, there is no regulatory requirement for IRBs to consider an investigator's financial conflict of interest, but they have the ultimate regulatory charge of protecting human subjects. It is understood that scientific objectivity requires honesty and integrity in reporting the data, the selection of research subjects, and the administration of proper informed consent. It is in this context that IRBs can take into consideration issues of conflict of interest.

As we have discussed in an earlier chapter, IRBs are appointed by research institutions to protect human research subjects by determining whether the research protocol should proceed or not. The members of academic IRBs are selected from employees of the research institution, with one member selected by the institution from the community as required by law (OPRR, 1993; 45 CFR 46; 21 CFR 56; 21 CFR 50). An important function of an academic institution is to seek funds from private as well as public sources. In order to obtain funding, principal investigators have a great interest in obtaining the approval of an IRB for their research protocols. This approval provides the institution and the investigators with funding to carry out the research. Therefore, funding satisfies both the institutions and the investigators in accomplishing their goal of growth, advancement, and, more importantly, the acquisition of new knowledge. In addition, a sponsor of research, if this is a private corporation, is under great pressure to produce profits for stockholders. The introduction of new drugs to the market is one of the most important methods for corporations to acquire profits. The interests of human subject volunteers may be compromised in such a hard pressed and financially rewarding situation for the investigator and the institution.

These facts did not escape the DHHS Inspector General's report on IRBs in 1998 (Brown, 1998a). The report states: "In an increasingly competitive research environment, however, in which those centers are seeking to maximize clinical research revenues, the

IRBs in these centers can also experience conflicting pressures." The report also recognizes the close personal and professional relationship between members of an IRB and their colleagues within the same institution. Thus, the report cites this close relationship as evidence of the compromised independence of IRBs.

Private for-profit IRBs experience conflict of interest because their own for-profit business demands continuous evaluation and approval of the large number of protocols before them. Private IRBs are a growing trend for clinical trials submitted to the FDA. The Office of Inspector General's report states that "for profit IRB might compromise its review process to advance the financial well being of the firm (Brown, 1998b)." These conflicting interests have therefore been handled by uncoupling the management of the private corporation from its IRB deliberations and decisions.

These conflicting interests make the nature of the function of private IRBs dramatically different from academically-based IRBs. Private IRBs, therefore, act not merely as gatekeepers of clinical trials to protect the public's welfare but as intermediary negotiators between sponsors and investigators, and the federal requirements in order to ensure that sponsors and investigators fully comply with federal regulations. It is in the interest of sponsors and investigators and the private IRBs not to run afoul of federal regulations. This role of private IRBs is similar to the public accounting firms that insure financial institutions comply with federal requirements. It is not in the interest of either party to bypass the federal regulations because public accounting firms' regulation and, in this case, their fiducial responsibility, can be in jeopardy, and the sponsor's and investigator's entire clinical program can be at risk.

The NIH, on June 5 2000, issued a Notice (OD-00-40), which is available on their website, entitled "Financial conflicts of interests and research objectivity: issues for investigators and institutional review boards." In this Notice, the NIH suggests that IRBs should consider some strategies to deal with conflict of interest of investigators. Among the strategies suggested are

1. Obtain information regarding how other IRBs deal with conflict of interest
2. Insure that IRBs are aware of institutional conflict of interest policies, if available, and that informed consent forms refer to that policy
3. See that the investigator and others directly involved in the research study complete a short-question questionnaire regarding conflict of interest

4. Insure that IRB members receive training and education in the responsible conduct of research and conflict of interest

FDA REQUIREMENTS FOR FINANCIAL DISCLOSURE BY CLINICAL INVESTIGATORS

Recently (21 CFR 54), the FDA began requiring that all clinical investigators should disclose their financial interests in the outcome of the clinical studies for submission for marketing. These disclosures are required for submission to the FDA with the marketing application of new human drug and biological products. The applicant is responsible for filing the appropriate papers. The applicant for new drug marketing will either file a disclosure form for all clinical investigators who participated in the study or a certificate attesting to the absence of any financial interest of the clinical investigator(s) in the outcome of the study. The financial disclosures are for the duration of the study and for 1 year after its completion. They also cover the spouse and each dependent child of the investigator. Disclosure or the certification covers, in part, the following financial interests

1. Equity interest in the sponsor or the study: ownership interest, stock options, and equity in publicly traded corporations that exceeds $50,000/yr
2. Propietary interest such as patents, trademarks, copyrights, or licensing agreements
3. Payment by the sponsor of the study exceeding $25,000 to the investigator and/or the investigator's institution to support the activities of the investigator

The applicant is also responsible for keeping records of information regarding financial disclosure, among which are complete records of any financial interests and arrangements, significant payments, and the financial interest held by the clinical investigator.

DHHS DRAFT INTERIM GUIDANCE

The DHHS (DHHS, 2000) issued draft interim guidance on "Financial relationships in clinical research: issues for institutions, clinical investigators and IRBs to consider when dealing with issues of financial interest and human subject protections." This guidance is meant to assist institutions in the implementation of the financial conflict of interest regulations of the NIH and the FDA. The draft addresses five players in this issue:

1. *The institution* It encourages that
   a. All institutions should have a conflict of interest policy and, if appropriate, establish a conflict of interest committee.
   b. IRBs should be free and autonomous to make decisions.
   c. Financial conflict of interest information should be collected from all IRB staff and the chairman.
   d. A training and education program that includes issues of conflict of interest should be established for clinical investigation and IRB members.
   At present, there is a great deal of debate regarding research institutions having equity in a corporation and at the same time conducting clinical trials for the same corporation.
2. *Clinical investigator* The guidance encourages clinical investigators to consider the potential effect of financial conflict of interest on the study, informed consent, and other relevant issues. Furthermore, it suggests that the conflict of interest committee or its equivalent should review the investigator's financial dealings with the sponsor.
3. *IRB members and staff* The guidance suggests that the IRB chairman should discuss with IRB members their potential conflict of interest and how this should be managed. IRB members should recuse themselves from deliberations when they have conflict of interest.
4. *IRB review of protocols and approval of consent documents* The guidance encourages all concerned parties that, in the presence of an actual conflict of interest that is problematic and not manageable, the institution should consider that the study should not go forward. In a lesser conflict of interest, a clear delineation of the mechanism to manage the conflict should be discussed and implemented.
5. *Consent* The guidance suggests that the consent document should consider including the source of funding and types of financial arrangement. Human research subjects should be informed of the conflict and how it is managed.

In response to a request from DHHS to comment on the DHHS draft, the National Human Research Protections Advisory Committee (NHRPAC) sent a detailed letter supporting and suggesting the expansion of the intent of DHHS to address issues of conflict of interest at all levels of the institutions, IRBs, and investigators.

NBAC'S REPORT AND CONFLICT OF INTEREST

The NBAC's report (NBAC, 1998) on decisionally impaired people recommended several steps to ensure the lessening of conflict of interest. Among the recommendations in the report are

1. All IRBs must include two members who suffer from the disorder under consideration.
2. The creation of a national Special Standing Panel (SSP) to recommend to the Secretary's approval of the proposed high-risk research protocols on decisionally impaired people that have no medical benefits to the subject. They recommended that the Panel should have diverse membership.
3. All IRBs require an independent, qualified professional to pass judgment on higher than minimal risk research protocols, and the capacity of patients to consent before enrollment.

These recommendations were all directed to lessen the conflict of interest of sponsors and investigators with the need to approve the project and to protect the patient's welfare. 45 CFR 46.107(e) forbids any IRB member to participate in the deliberation of any project in which the member has a conflict of interest. The OPRR, for example has cited cases of this nature as a cause of their action against a research institution (OPRR, 1999). An IRB member who is also the Director of the Office of Research, whose job is to solicit funds from similar outside sources, illustrates a conflict of interest. Another example would be if an IRB member is a direct employee of the investigator of the project under consideration. It is prudent for IRBs to avoid such obvious causes of conflict of interest.

Quiz: Choose the Best Response

1. Conflict of interest can be characterized by:
   a. Two competing interests of the investigator
   b. Conflict between private good and public good
   c. Nonmeritorious influence on the decision maker
   d. All of the above
   e. None of the above
2. Conflict of interest can influence the following steps of research, *except:*

a. Gender of the investigator
b. Selection of subjects
c. Design of the protocol
d. Analysis of data
e. Publication

3. Disclosure of financial conflict of interest in clinical trials is required by:
a. The Common Rule
b. The FDA
c. The NIH
d. a, b, and c
e. None of the above

4. Private IRBs differ from research institute IRBs by the following:
a. Accepted by the FDA
b. Review FDA-bound clinical trials protocols
c. Conduct research studies
d. Not required to have a community member
e. a and b

5. The FDA requires financial disclosures such as, *except:*
a. Payment by the sponsor exceeding $50,000 to the investigator or the institution
b. Equity interest exceeding $50,000
c. Payment by the sponsor exceeding $25,000 to the investigator or the institution
d. Propietary interest such as patents
e. None of the above

Case Studies

Case 1

A physician/investigator receives $1000 per patient enrolled in research in an IND approved protocol from a pharmaceutical company. The protocol calls for strict inclusion/exclusion criteria for cancer patient enrollment in this study.

- Should the protocol mention the payment and the amount?
- Should the patient be informed and why?
- What questions should the IRB ask?
- How do you manage this conflict of interest – if you think there is one?

Case 2
A pharmaceutical company pays an investigator $15,000 per year for consultation. The investigator is a university researcher and is conducting research on human subjects sponsored by the NIH. The NIH study is on a drug that is on the market but the study is for treatment (yet unapproved) for a different disorder. The drug is marketed by the pharmaceutical company.

- Is there a potential conflict of interest?
- Should the investigator disclose the relationship and to whom?
- How should the conflict be managed if one exists?

Case 3
An investigator owns a small start-up biotechnology company and is a university professor. He has 15% equity in the biotechnology company. The university has 5% equity in the company. The company has one drug for which the clinical trials are conducted by the investigator at the company facilities. The company pays the professor 30% of his salary and the university pays the other 70% of it.

- How many types of conflict of interest exist?
- What potential conflict of interest exists?
- How should the conflict of interest be managed?

References

Brown JG (1998a) Department of Health and Human Services, Office of Inspector General, on: Institutional Review Boards: a Time for Reform (OEI-01-97-00193).

Brown JG (1998b) Department of Health and Human Services, Office of Inspector General on: Institutional Review Boards: the Emergence of Independence Boards (OEI-01-97-00192).

21 Code of Federal Regulations parts 50, 54, 56

45 Code of Federal Regulations part 46

Department of Health and Human Services (DHHS) (2000) Draft Interim Guidance – Financial Relationships in Clinical Research: Issues for Institutions, Clinical Investigators, and IRBs to Consider When Dealing With Issues of Financial Interest and Human Subject Protection. Available at: http://ohrp.osophs.dhhs.gov/humansubjects/finreltn/finmain.htm

Garner BA ed. (1999) Black's Law Dictionary, 7th edn. St Paul, Minnesota: West Group.

National Bioethics Advisory Commission (NBAC) (1998) Research Involving Persons with Mental Disorders That May Affect Decision-making Capacity, vol. I: Report and Recommendation. Rockville, Maryland: NBAC. Available at: http://www.bioethics.gov

Office of Protection from Research Risks (OPRR) (1999) OPRR Compliance Activities: Common Findings and Guidance 11/29/99. Available at: http://ohrp.osophs.dhhs.gov/references/findings.pdf

Office of Protection from Research Risks (OPRR) (1993) Protecting Human Subjects: Institutional Review Board Guidebook. Washington, DC: US Government Printing Office.

Porter D (1992) Science, scientific motivation, and conflict of interest in research. In Ethical Issues in Research (D Cheny ed.) pp 114–125. Frederick, Maryland: University Press.

Shamoo AE & Dunigan C (2000) Introduction to ethics in research. *Exp. Biol. Med.* 224, 205–210.

Chapter 7

The Use of Decisionally Impaired People in Research

The Nuremberg Code and all of the US commissions on ethical issues concerning human research subjects, as well as federal regulations, require that those who sign an informed consent document must have the capacity to consent. Therefore patients with serious mental disorders and dementia pose a special problem for their enrollment in research. The elements of informed consent, especially comprehension, lack of duress and coercion, and providing sufficient information, become particularly crucial for this population. Recently there has been criticism of the current standards of enrollment of such groups in research. The NBAC has recommended additional protections for these vulnerable people. Among their recommendations are increased membership of IRBs for this population, their families or their advocates; assessment by an independent professional of their capacity to consent; and the creation of a national Special Standing Panel (SSP) to pass judgment on enrolling patients in high-risk research that has no medical benefit for them.

Learning Objectives

1. The need for special protections for decisionally impaired people
2. The association of decisional impairment with serious mental illness and dementia
3. NBAC-recommended additional protections such as increased membership of IRBs for this population, their families or their advocates; and assessment of their capacity by independent professionals

The special regulations involving vulnerable research subjects derive moral authority from the Nuremberg Code, formulated in 1947, and the recommendations of the National Commission in the late 1970s. The Nuremberg Code's first enduring principle contains within it that "the person involved should have legal capacity to give consent." This statement clearly bars decisionally impaired individuals who have no legal capacity to consent, as well as children, from signing an informed consent form.

The Code also states that the person "Should be so situated as to be able to exercise free power of choice, without the intervention of any element of force, fraud, deceit, duress, overreaching, or other ulterior form of constraint or coercion." This statement identifies those individuals who are without freedom because of being confined in an institution or in a prison or other facilities as not being able to give meaningful informed consent.

The Code goes on to state that the person "Should have sufficient knowledge and comprehension of the elements of the subject matter involved as to enable him to make an understanding and enlightened decision." This statement identifies those with decisional impairment (i.e. those with serious mental illness) who may not be able to give valid or meaningful informed consent (45 CFR 46; OPRR, 1993). In addition, young children lack the ability to comprehend the risks and benefits of enrolling in research, as will be discussed in Chapter 8.

The ethical principle of "respect for the person" enunciated by the National Commission and the DHHS regulations (i.e. the Common Rule) provide authority to prohibit the enrollment of any human research subject without "voluntary consent" (ACHRE, 1995).

In this chapter we will discuss a special class of research subjects that the federal regulations consider to be particularly vulnerable, namely, those who are decisionally impaired.

The National Commission first officially recognized and included "mental disability" as one of the categories of vulnerable people for the purpose of enrollment as research subjects. The Commission listed mentally disabled people, together with children and prisoners, as the three most vulnerable groups, issuing specific reports for additional protections for children and prisoners but not those who are mentally ill. The vulnerable status of these people was further emphasized by the President's Commission for the Study of Ethical Problems in Medicine and Biomedical and Behavioral Research (1983), following the earlier recommendation by National Commission Belmont Report (National Commission, 1978; ACHRE, 1995).

The NBAC (NBAC, 1998) used the term "decisionally impaired" instead of "mentally disabled" as used earlier. This was meant to emphasize that decisional impairment is the operating phrase for those with mental disorders, especially those with severe and chronic mental disorders, rather than mental disability *per se*. We will use both terms interchangeably to indicate that a number of such individuals may not comprehend the risks and benefits of participating in a research protocol as human subjects. This impairment becomes more compromised when decisionally impaired persons are institutionalized, voluntarily or involuntarily, in confined environments.

It is important to note that the presumption should be that mentally disabled people are capable of comprehension unless they lack this faculty and are thus decisionally impaired. The percentage of mentally ill people who are decisionally impaired varies among the types of mental disability. The more severe and chronic the mental disorder, the more likely that the person suffers from impairment of comprehension and decisional capacity. Thus, those patients with schizophrenia, manic depression, major depression, or anxiety disorders may be the most likely to have problems with comprehension. Other mental illnesses may also impair comprehension, however. In addition, the effect of mental disorders on comprehension may be intermittent. Therefore, comprehension status needs to be re-evaluated periodically (Rabins, 2002). The frequency of comprehension evaluation during the initial and ongoing consent process can be incorporated into the research protocol.

The National Commission was explicit in providing additional protections for those with mental disability.

> "Special provision may need to be made when comprehension is severely limited—for example, by conditions of immaturity or mental disability. Each class of subjects that one might consider as incompetent (e.g., infants and young children, mentally disabled patients, the terminally ill and the comatose) should be considered on its own terms. Respect for persons also requires seeking the permission of other parties in order to protect the subjects from harm. Such persons are thus respected both by acknowledging their own wishes and by the use of third parties to protect them from harm. The third parties chosen should be those who are most likely to

understand the incompetent subject's situation and to act in that person's best interest. The person authorized to act on behalf of the subject should be given an opportunity to observe the research as it proceeds in order to be able to withdraw the subject from the research, if such action appears in the subject's best interest!" (National Commission, 1978, p 6).

The President's Commission expressed support for the National Commission's recommendation regarding the mentally disabled people.

> "The National Commission's recommendations on research involving children and the mentally disabled should be acted upon promptly. Ethical concerns about these individuals revolve around the issue of informed consent. In order for research on the causes, treatment, and prevention of pediatric diseases and of emotional and cognitive disorders to proceed in an ethically acceptable manner, the National Commission had urged the adoption of specific protections for children and the mentally disabled" (President's Commission, 1983, p 54).

Despite the recommendations of two important Commissions, the federal regulations on mentally disabled people were never issued as a separate subpart of 45 CFR 46. However, the federal regulations currently in effect (45 CFR 46 became the Common Rule (1991)) discuss added protections for the mentally disabled in regard to IRB composition.

> "If an IRB regularly reviews research that involves a vulnerable category of subjects, such as children, prisoners, pregnant women, or handicapped or mentally disabled persons, consideration shall be given to the inclusion of one or more individuals who are knowledgeable about and experienced in working with these subjects" (Code of Federal Regulations 45 CFR 46, 1991, p 7).

The interpretation and the spirit of the Common Rule with regard to the mentally disabled people have been more cautious and strict in the recent past than their interpretation in the 1970s and 1980s. The former OPRR's institutional review board guidebook (1993) emphasizes that

> "research involving persons whose autonomy is compromised by disability or restraints on their personal freedom

> should bear some direct relationship to their condition or circumstances. Persons who are institutionalized, particularly if disabled, should not be chosen for studies that bear no relation to their situation just because it would be convenient for the researcher."

Scholars, researchers, advocacy groups, IRBs, and the OPRR are not all in agreement about what constitutes an acceptable degree of risk when mentally disabled individuals are enrolled in research programmes (Shamoo, 2002). IRBs are now left to their own interpretation of what constitutes decisional impairment and what is greater than minimal risk. However, thay are duty bound to insure that the research protocol does not place mentally disabled people at undue risk of injury.

This earlier lack of attention to clarity in federal regulations regarding added protections for mentally disabled persons might have sewn the seeds of controversy 10 years later. Since the early 1990s there has been increased criticism of how mentally disabled people are recruited and enrolled in research experiments (Shamoo and O'Sullivan, 1998). This criticism may have contributed to the motivation of the former federal OPRR and the FDA to investigate and temporarily halt research operations in several major institutions nationwide (Marshal, 1999).

THE NBAC REPORT

The NBAC has addressed these issues as one of their prime concerns. This Commission was appointed in 1995 by the President to address pressing issues confronting our society by the major advances in biology. For over 2 years the NBAC studied and held hearings on the issue of the use of decisionally impaired participants in research, and published their report in 1998, entitled: "Research involving persons with mental disorders that may affect decisionmaking capacity." In its executive summary it states that

> "A cogent case can be made for requiring additional special protections in research involving as subjects persons with impaired decision making capacity, but has chosen to focus this report on persons with mental disorders, in part because of this population's difficult history of involvement in medical research (NBAC, 1998, p ii)."

The NBAC advises that its recommendation should be accomplished by "the creation of a new subpart in 45 CFR 46."

Recognizing the fact that commission recommendations are not immediately promulgated into regulations and implemented, the NBAC advised IRBs and institutions to adopt and comply voluntarily with their recommendations until new regulations are written. Even though the commission's report carries no power of law, it does carry with it the moral authority of a presidentially appointed commission representing a cross-section of our society. Highlights of NBAC's recommendations are

1. IRBs involved in reviewing protocols concerned with decisionally impaired populations should contain "at least two members who are familiar with the nature of these disorders and with the concerns of the population being studied."
2. "For research protocols that present greater than minimal risk, an IRB should require that an independent, qualified professional assess the potential subject's capacity to consent."
3. For research that involves greater than minimal risk and offers no prospect of direct medical benefit to subjects, the protocol should be referred to the national SSP for decision. The exception is when the subject has a valid legally authorized representative to act on his or her behalf.

COGNITIVE IMPAIRMENT

Cognitive impairment manifests itself because of numerous causes. Among them are psychiatric disorder, dementia, mental retardation, use of drugs and alcohol, severe illness, and disability. Even though the formal legal term is "legal capacity," cognitive impairment is its actual basis. The person who lacks capacity to consent cannot weigh the pros and cons of enrolling in a research project, or appreciate the consequences of enrollment, and is thus unable to make a true choice. For the sake of this discussion, the term decisional impairment is the more appropriate one.

A person may be decisionally impaired concerning making a decision regarding enrollment in a research project but at the same time may not be cognitively impaired. There are two broad categories associated with decisional impairment. The first is characterized by the fact that cognitive impairment is the primary manifestation of disease (e.g. delirium and dementia such as in Alzheimer's disease). The second is characterized by the fact that cognition is not central to the manifestation of the disease (Rabins, 2001) (e.g. schizophrenia, and all the various mood disorders). In these disorders there are no available data about what percentage of those affected are expected to have permanent or intermittent cognitive impairment. It is therefore prudent to have an independent capacity assessment of decision making ability prior to enrollment.

The recognition of decisional impairment by the investigator and the IRB is the most relevant issue in protection of the interests of human research subjects (Rabins, 2002). In addition, the National Commission's recommendations and the federal regulations designate institutionally infirm persons as the most vulnerable to a compromised ability to exercise free choice in addition to problems with comprehension of the risks and benefits.

ENROLLMENT OF DECISIONALLY IMPAIRED PEOPLE

In order for investigators and IRBs to conduct research with populations who may be decisionally impaired, they must not only be aware of the current federal regulations but also of recent developments in insuring the protection of this population. Among these are that (1) IRBs should have one or two members of the studied group; (2) the research in question must be relevant to the patient's disorder; (3) there should be adequate and independent evaluation of a participant's comprehension to consent; and (4) the investigator should have considered the risks to the patient and whether they are outweighed by the benefit to the patient and/or generalizable knowledge. Furthermore, the investigator should state how he or she would deal with and remedy these risks to the patient during the research. After weighing these factors, the investigator should describe in detail the method of patient recruitment as part of the submission of information to the IRB.

Decisional impairment may also be intermittent. In addition, it may become florid during the subject's enrollment in the research protocol. Researchers should anticipate some of these problems and have a plan of action to insure the safety of patients.

Quiz: Choose the Best Response

1. In signing the informed consent form, human volunteers' comprehension is part of the:

a. Nuremberg Code
b. National Commission report
c. Common Rule
d. President's Commission report
e. All of the above

2. Current federal regulations require all IRBs to have:
a. Two members of the vulnerable group present during their deliberations
b. One or more individuals who are members of the group enrolled in the research study
c. One or more individuals who are knowledgeable about and experienced in working with this group
d. No individual from the research subjects
e. None of the above

3. Vulnerable human subject categories as identified by 45 CFR 46 are:
a. Children, pregnant women, handicapped people, and mentally ill people
b. Children, prisoners, pregnant women, handicapped people and mentally disabled people
c. Children, pregnant women, and prisoners
d. Pregnant women, mentally ill people, and children
e. None of the above

4. The federal regulations provide special subparts in 45 CFR 46 for the following vulnerable groups:
a. Children, pregnant women, and prisoners
b. Children, prisoners, pregnant women, and mentally ill people
c. Pregnant women, mentally ill people, and children
d. Children and prisoners
e. None of the above

5. The NBAC's 1998 report recommends that decisionally impaired people should be given the following added protections:
a. IRBs should have at least two members who are familiar with the disorders of the population studied.
b. Each decisionally impaired participant carries accident insurance paid for by the sponsor.
c. Research protocols that present greater than minimal risk should include capacity assessment by an independent qualified professional.

d. a, b, and c are correct
e. None of the above

Case Studies

Case 1
A protocol calls for 50 schizophrenia patients to be enrolled in a study to test a new promising antipsychotic medication. It also calls for half of the enrolled patients to be on placebo and the other half on the new experimental drug. The investigator plans to recruit some of his own patients and others from the community.

- What precautions, if any, the protocol should undertake to protect the research subjects?
- What should the informed consent document contain?
- Who should sign the informed consent document and why?

Case 2
Twelve manic-depressive patients were washed out of their medication for 2 weeks. The experimental protocol calls for subjecting the patients to a positron emission tomographic scan in order to understand the effect of the new drug on receptor binding.

- What questions should the patient ask?
- Can these experiment be approved and why?
- What safeguards, if any, should be in place?

Case 3
A researcher in a university hospital is conducting research for a pharmaceutical company. The protocol calls for 300 persons diagnosed with manic depression. The researcher intends to recruit all the subjects from his own clinic and the hospital. The study entails testing a new drug.

- What else should the patient be told?
- What information should be included in the informed consent document?
- Who should sign the informed consent document?

Case 4
A researcher at a private research institution is proposing to conduct a survey of home sheltered patients. The research institution runs the home shelter. The study is to find out who is complying with the medication schedule.

- Is there any problem with this protocol?
- Do you think the study should be redesigned and how?
- What should you tell the patient?
- What should be included in the protocol?

References

Advisory Committee on Human Radiation Experiments (ACHRE) (1995) Final Report, stock number 061-000-00-848-9. Available from: Superintendent of Documents, US Government Printing Office, Washington, DC. Tel: (202) 512-1800; Fax: (202) 512-2250.

45 Code of Federal Regulations part 46 (as of 1991, also called the Common Rule).

Marshal E (1999) NIMH to screen studies for science and human risks *Science* 283, 464–465.

National Bioethics Advisory Commission (NBAC) (1998) Research Involving Persons with Mental Disorders That May Affect Decision-making Capacity, vol. I: Report and Recommendation. Rockville, Maryland: NBAC. Available at: http://www.bioethics.gov

National Commission for the Protection of Human Subjects of Biomedical and Behavioral Research (1978) Belmont Report: Ethical Principles and Guidelines for the Protection of Human Subjects of Research. Washington, DC: US Department of Health, Education, and Welfare.

Office of Protection from Research Risks (OPRR) (1993) Protecting Human Subjects: Institutional Review Board Guidebook. Washington, DC: US Government Printing Office.

President's Commission for the Study of Ethical Problems in Medicine and Biomedical and Behavioral Research (1983) Summing up, Washington, DC: US Government Printing Office.

Rabins PV (2002) Mental disorders affecting decisional capacity. In: Research and Decisional Capacity (AE Shamoo ed.) London, UK: Taylor & Francis Group, in press.

Shamoo AE (ed.) (2002) Research and Decisional Capacity London. UK: Taylor & Francis Group, in press.

Shamoo AE & O' Sullivan J (1998) The Ethics of Research on the Mentally Disabled. Chapter 21 in: Health Care Ethics – Critical Issues for the 21st Century (JF Monagle & DC Thomasma eds.), pp 239–250 Maryland, Gaithersburg: Aspen Publishers.

Chapter 8

The Use of Children in Research

Children, because of their age and reduced capacity to comprehend, lack the legal capacity to consent for enrollment in research as human subjects. Both the Nuremberg Code and the National Commission recognized children's limitations and provided for their greater protection compared with adults when enrolled in a research project. Children are named in 45 CRF 46 as one of the vulnerable groups requiring additional protection, such that parents or legally authorized persons should sign the informed consent document on behalf of the child. It further requires that research protocols with children should be divided into four categories based on the degree of risk to the child. The regulations prohibit the use of children in research that has greater than minimal risk and no prospect of direct medical benefit, except with the deliberation and approval of the Secretary of the DHHS. Lower categories of risk require careful balancing of the risks and benefits and the assent (if possible) of the child.

Learning Objectives

1. Why are children a vulnerable group?
2. How should IRBs deal with research protocols involving children?
3. What are the levels of risk of research protocols?
4. When should the IRB refer the research protocol for approval to the Secretary of the DHHS?

The Nuremberg Code's first enduring principle contains the statement that "...the person involved should have legal capacity to give consent." This statement clearly restricts allowing children below the legal age and minors to sign informed consent documents. The Code further states that the person "Should have sufficient knowledge and comprehension of the elements of the subject matter involved as to enable him to make an understanding and enlightened decision" (ACHRE, 1995). This statement prohibits children, because of their lack of ability to comprehend the risks and benefits, from enrolling, on their own, as research subjects.

Additionally, the ethical principle enunciated by the National Commission, of "respect for person," and the DHHS's regulations (i.e. the Common Rule) provided the legal authority to prohibit the enrollment of any human subject without "voluntary consent."

Federal regulation 45 CFR part 46 subpart D provides for "additional protections for children" when enrolled as research subjects. The primary role of an IRB is to protect the public (i.e. human subjects) from undue risk and harm. Children, therefore, represent a special burden to IRBs and thus require their special attention. As a class they lack the autonomy, and young children in particular lack the requisite comprehension of risks and benefits, to give valid consent. Therefore, federal regulations recognize the limitations of children and require that "proxy consent" be used instead. Even though the regulations do not use this commonly used term, they mention consent by a parent, guardian, or legally authorized representative. In addition, the child must give his or her assent with an affirmative sign of approval of participating in the research.

NEW FDA MANDATE

The majority of drugs used to treat children have not been tested in children. The pharmaceutical industry is reluctant, owing to liability concerns, to test drugs on children. Therefore, children represent "therapeutic orphans." The administration of research drugs to children has therefore been as "off label" use. The ethical dilemma arises when the interests of children as a group collide with the interests of those children who are used as subjects to test new drugs that have potential risks and no benefit (for reviews see Tauer, 1999; Grodin and Glantz, 1994). Moreover, off label use carries a large risk to children. These ethical issues became more focused when President Clinton, in 1997, ordered changes in the testing of drugs that are to be used by children. This was followed, in 1998, by the FDA issuing a ruling mandating that the pharmaceutical industry must test drugs and biological products in children if they are to be used in children. The logic behind these changes was the same as that requiring the inclusion of women and minority groups in research trials. The burden is now on the investigator to explain why children are excluded when an IND application is submitted. There is a great deal of debate among government officials, congress, media, and advocacy groups with regard to when and how children ought to participate in research. Examples of the discussion topics are the definition of minimal risk; the definition of assent; safeguards; the differing maturity level of a child; and the most hotly debated issue of what constitutes a direct benefit to the child.

CATEGORIES OF RISKS

According to federal regulations (45 CFR Part 46), IRBs must classify proposed research on children into one of four categories. This classification is based on an analysis of probable risks, possible benefits, and associated discomforts. The analysis of possible risk and discomfort to the child should be weighed against the expected direct benefit to the child, class of subjects of the child's, and to the society at large (OPRR, 1993).

After careful deliberation and thoroughly weighing the degrees of risks and benefits, and the study design, the IRB assigns one of the four categories that will play an important role in its approval. These categories are primarily based on the degree of risk to the child.

1. *Minimal risk research* This category is defined as "the probability and magnitude of harm or discomfort anticipated in the proposed research is not greater, in and of itself, than ordinarily encountered in daily life or during the performance of routine physical or psychological examinations and tests (45 CFR 46, Section 102 (i))." This definition of minimal risk is the same for all human subjects, whether children or adults. The National Commission has mentioned examples of minimal risk research protocols for children, such as surveys, noninvasive

physiological monitoring, routine immunization, and obtaining blood and urine samples. This category, if it meets all other conditions, can be approved by the IRB.

2. *Greater than minimal risk research with the prospect of direct medical benefit to the child* The IRB must determine
   a. If such risk is justified by the anticipated benefit to the subject
   b. If the risks/benefit relationship to the subject is as favorable as for alternative approaches
   c. The assent of the subject

   The IRB must examine carefully all research protocols that include the use of placebos. If the protocol involves greater than minimal risk to the subject and has no direct medical benefit, the placebo arm, according to Levine (1988, p 251), may or may not be approvable. The practice has been to approve such protocols with the assumption that the benefit to the class of disease would have direct benefit to the child. However, literal reading of the regulations leaves little room for such broad interpretation. It is clear that there is a great deal of controversy regarding placebo control groups as part of research protocols for children and whether or not they should be approved by IRBs.
3. *Greater than minimal risk with no prospect of direct medical benefit to the child but likely to produce generalizable knowledge about the subject's disorder or condition* This is an approvable category if
   a. The risk is a minor increment over minimal risk
   b. The inherent research protocol is within the medical and other procedures that the subject may experience
   c. The procedure produces generalizable knowledge
   d. The assent of the child is secured

   The federal regulations left the definition and application of "minor increase over minimal risk" to each IRB. This has led to great variance of interpretation. Again, the issue of a placebo arm remains unclear.
4. *Greater than minor increase over minimal risk with no prospect of direct medical benefits to the child* This is an unapprovable category of research with children. However, under certain circumstances and after referral to the DHHS, the Secretary may approve the protocol if the research

> outcome would help to understand and contribute to the general welfare of children. The DHHS Secretary would give approval only after consulting experts in the field and soliciting public comments on the protocol.
>
> Under this category, children who are wards of the state in any capacity can participate if the research is related to their status as wards or the research is conducted in educational and medical settings; and if each child has an appointed advocate.

Institutional assurances for research protocols involving children require that each institution should protect the children's rights and welfare regardless of the source of funding. The OHRP encourages institutions to appoint a panel of experts similar to those that the Secretary may appoint to review certain research on children. When a child is a ward of the state, the child's appointed advocate has the duty and authority to act in the child's best interest when he or she is asked to enroll in a research project.

IRBs have an important and difficult task to assess all of the risk factors to children. Some of the questions the members of an IRB should ask are

1. What is the meaning of minimal risk?
2. How are the risks assessed for children in the protocol?
3. What is meant by "minor increase" over minimal risk?
4. Are pain and discomfort considered well in the protocol?
5. How does the protocol minimize risks?
6. Is the age of a child an important factor in experiencing pain, discomfort, and risk?
7. Is there compensation for the pain, discomfort, and risk?

CHILD'S ASSENT

In practice, no assent forms are needed for children aged below 7 years. However, their agreement should be sought in a manner that is not coercive. For example, the child could be told that no one will be annoyed if he or she refuses to participate. For those aged between 7 and 12 years, a separate form is advisable. However, adolescents may not need a separate form. The child must be able to understand the assent language, which should consist of simple words and concepts.

RESEARCH IN SCHOOLS

The reader should consult individuals who are familiar with the conduct of research in schools. This type of research is governed by two laws: the Family Educational Rights and Privacy Act (FERP), and the Protection of Pupil Rights Amendment (PPRA). The PPRA regulates survey research in schools, defines the categories of research that be conducted in schools, and defines the rights of parents.

Quiz: Choose the Best Response

1. Children cannot consent to enroll in a research protocol because:
   a. They lack comprehension of the risks and benefits
   b. They are below 18 years of age
   c. 45 CFR 46 prohibits it
   d. It is consistent with the Nuremberg Code
   e. All of the above
2. Proxy consent can be given for the child by:
   a. Parent
   b. Guardian
   c. Legally authorized person
   d. a, b, and c are correct
   e. None of the above
3. The logic behind the new FDA mandate requiring testing new drugs on children is the following, *except:*
   a. Risk of harm because of off label use
   b. Children's body weight is sufficient to calculate dosage of the drug
   c. Greater economic liability to industry when children are used as subjects
   d. Development of drugs to children as a group
   e. None of the above

Note: The four categories of risk in research with children are
   1. Minimal risk
   2. Greater than minimal risk with the prospect of direct medical benefit to the child but likely to produce generalizable knowledge about the child's disorder
   3. Greater than minimal risk with no prospect of direct medical benefit to the child but likely to produce

generalizable knowledge about the child's disorder or condition
4. Greater than minor increase over minimal risk with no prospect of direct medical benefit to the child

4. Category (1) below is approvable if:
a. The probability and magnitude of harm or discomfort is not greater than that ordinarily encountered in daily life or during these tests
b. The probability and magnitude of harm or discomfort is less than that ordinarily encountered during these tests
c. The probability and magnitude of harm or discomfort is minor above that ordinarily encountered in daily life
d. a and c are not correct
e. None of the above

5. Category (2) below is approvable if one of the following is met:
a. There is assent of the subject and consent of parent or guardian
b. There is assent of the subject, the risk – benefit relationship to the subject is as favorable as to the alternative approaches, and the risks are justified by the benefit to the subject.
c. There is consent of a parent as well as a favorable risk – benefit ratio
d. a and c are correct
e. None of the above

6. Category (3) below is approvable if only:
a. The assent of the child is secured
b. The knowledge is generalizable
c. The risk is a minor increment over minimal risk
d. The research procedure is within acceptable medical experience
e. All of the above

7. Category (4) below is approvable if:
a. The Secretary of DHHS approves it
b. The IRB and the Secretary approve it
c. The IRB approves it
d. The Secretary of the DHHS approves it after consultation with a panel of experts and after soliciting public comments on the protocol
e. The Secretary of the DHHS approves it after consultation with a panel of experts

Case Studies

Case 1

A research protocol calls for using 60 healthy children (6–18 years old) in a study to test the frequency of children contracting influenza. The protocol calls for the participants to undergo an initial physical examination, monthly visits to the investigator, and filling in a detailed survey questionnaire.

- What should the category of this experiment be and why?
- What additional information should be included in the protocol?
- What information should the informed consent form include?

Case 2

A research protocol calls for the use of 100 children with cancer in a study to test a new drug. The preliminary evidence indicates that the new drug is superior to the current standard of care medication.

- What should the category of this experiment be?
- What additional information should the parents be given?
- What safeguards should be in place?

Case 3

A new vaccine is being tested on children. A third of 300 children will receive the new vaccine, a third will receive the old vaccine, and a third will receive placebo.

- What category would you use to classify this experiment?
- Do you have any problems with the design and, if any, how would you change it?
- What should the parents be told?

References

Advisory Committee on Human Radiation Experiments (ACHRE) (1995) Final Report, stock number 061-000-00-848-9. Available from: Superintendent of Documents, US Government Printing Office, Washington, DC. Tel: (202) 512-1800; Fax: (202) 512-2250.

21 Code of Federal Regulations parts 50, 54, 56, 312, 314

45 Code of Federal Regulations part 46

Grodin MA & Glantz LH (eds) (1994) Children as Research Subjects – Science, Ethics, and Law. New York: Oxford University Press.

Levine RJ (1988) Ethics and Regulation of Clinical Research, 2nd edn. New Haven, Connecticut: Yale University Press.

Office of Protection from Research Risks (OPRR) (1993) Protecting Human Subjects: Institutional Review Board Guidebook. Washington, DC: US Government Printing Office.

Tauer CA (1999) Testing Drugs in Pediatric Populations: the FDA mandate. *Accountability Res.* 7, 37–58.

Chapter 9

The Use of Prisoners in Research

Prisoners, because of their incarceration, lack some of the prerequisites for informed consent for enrollment in research. The National Commission and federal regulations recognized prisoners' lack of free power of choice and provided for their greater protection when enrolled in research than those outside the prison walls. The added protections require that research on prisoners should have minimal risk related to their incarceration; be related to prison life and conditions; or have direct medical benefit.

Learning Objectives

1. Why are prisoners a vulnerable group?
2. How should IRBs deal with research protocols involving prisoners?
3. What are the special protections provided to prisoners volunteering for research?

We have in the previous two chapters discussed the use of two vulnerable groups: those who are decisionally impaired and children. Other vulnerable groups are identified in the federal regulations, such as pregnant women and prisoners. IRBs may also identify vulnerable groups that are not specifically identified in the regulations. In this chapter we will discuss the use of prisoners in research. As we stated earlier, the relevant section of the Nuremberg Code states that the person "Should be so situated as to be able to exercise free power of choice, without the intervention of any element of force, fraud, deceit, duress, overreaching, or other ulterior form of constraint or coercion." This statement identifies those individuals who are without freedom because of incarceration in a prison or other facilities as not able to give meaningful informed consent (ACHRE, 1995).

The use of prisoners in research was common prior to the report issued by the National Commission in the late 1970s. The primary ethical principle concerned in restricting the use of prisoners as research subjects is that they are in a compromised situation when it comes to the exercise of "free power of choice." In the USA there are over 2 million prisoners, largely from minority groups, so their use as research participants has the added dimension of a human rights issue. Prisoners are held under isolated conditions that could easily be construed as intimidating and coercive. Furthermore, the fact that the majority of prisoners are poor and not highly educated makes them vulnerable. It is for these and other reasons that the National Commission recommended and the federal regulations reflect special restrictions in the recruitment and use of this population in research prisoners.

Opposite to rights of autonomy are the benefits incurred to subjects when enrolling in a research project. The question arises of why prisoners should be denied the benefits of participating in research. Participants usually are compensated for their discomfort and their time, but in the case of prisoners the normal compensation package would be viewed as unduly coercive and exploitative. Moreover, time spent as participants in research may be a welcome "vacation" for prisoners from the daily boredom and inadequate conditions of prison life. One can ask why prisoners should not be paid the same amount as nonprisoners.

Another dimension is that maintaining confidentiality in a prison environment for those participating in research is practically impossible. Therefore, insuring parity of prisoners' pay with that of nonprisoners creates a new set of ethical problems. In one example of the truly exploitative use of prisoners as human subjects in research, Hornblum (1998) chronicled in his book, *Acres of Skin,* how, in the 1960s and 1970s, researchers used the skin on the backs of prisoners to test numerous drugs and perfumes for toxicity and carcinogenicity. These experiments were conducted at Holmesburg Prison in Philadelphia, Pennsylvania, by University of Pennsylvania researchers. Apparently, most of the records of these experiments were destroyed. All of the ethical principles mentioned above were severely tested under these conditions, such as the amount of payment and the degree of coercion; housing for those taking part in the experiment was better than that provided for other prisoners; human contact during the experiments may have been unduly coercive; and informed consent was barely informative.

At present, all federally supported research with human subjects must comply with the 45 CFR 46 (Common Rule). In the case of clinical trials bound to the FDA for drug approval, they must comply with the FDA equivalence to the Common Rule mentioned in earlier chapters. In addition, as with children, there are *additional* safeguards for prisoners when enrolled as research subjects. The most severe restriction is the fact that prisoners can be used primarily for behavioral research. Even with this restriction, the research protocol may involve prisoners only if (1) the research is of minimal risk related to the "causes, effects, and process of incarceration;" or (2) it is a study of the prison and its inmates as an institution; or (3) the research concerns conditions of prisoners as a class; or (4) it is therapeutic research with potential benefit to the prisoner. If the research is nontherapeutic, the Secretary can approve such research after consultation with experts and publicize the intent to approve such research.

There are further restrictions on using prisoners in research such as the fact that recruitment should be fair and not influenced by the prison's administration or a promise of leniency during parole proceedings. There is disagreement among scholars of how one defines "minimal risk" for prisoners. Surely, prisoners encounter different risks during their daily lives compared with nonprisoners. Therefore, some would argue that the standards of risk "normally encountered in daily lives" should be the same as everyone else outside of prison. The follow-up care of research subjects should be considered if needed.

The Federal Bureau of Prisons, for its own inmates, places further specific restrictions on prisoners' enrollment in research. In reality only minimal risk research will be approved by the Bureau for those under their jurisdiction.

Case Studies

Case 1

A research study conducted on prison inmates entails testing new antismoking medication. Most inmates are smokers. The study calls for the enrollment of 200 inmates. Half of the enrollees will be commenced on the new medication and the other half will receive a placebo.

- Is this an appropriate study design when carrying out research on prisoners?
- What should the prisoners be told and why?
- What questions should the IRB ask?

Case 2

Research is to be conducted on incarcerated drug addicts. The protocol calls for testing the effectiveness of a specific drug and the behavioral treatment of addicts.

- Should the IRB have a member who is knowledgeable about prisoners?
- What should the IRB ask?
- What should the informed consent document contain and why?

References

Advisory Committee on Human Radiation Experiments (ACHRE) (1995) Final Report, stock number 061-000-00-848-9. Available from: Superintendent of Documents, US Government Printing Office, Washington, DC. Tel: (202) 512-1800; Fax: (202) 512-2250.

Hornblum AM (1998) Acres of Skin. New York: Routledge.

National Commission for the Protection of Human Subjects of Biomedical and Behavioral Research (1979) Belmont Report: Ethical Principles and Guidelines for the Protection of Human Subjects of Research. Washington, DC: US Department of Health, Education, and Welfare.

Chapter 10

Process of Obtaining Approval for the Use of Human Subjects in Research

Recent actions taken by the OHRP and the unpleasant media coverage of these actions and other events have brought some of the frailties of IRB processes into question. The issues include an improper quorum; a lack of reporting of AEs; deficiencies in the informed consent process; and, in rare instances, fraud. In order for IRBs to carry out their duties, they need to collect appropriate and comprehensive information regarding proposed studies using human subjects. Investigators have an ethical and legal responsibility to comply with federal regulations. FDA regulations clearly state that investigators risk criminal prosecution in the event of willful false statements to the FDA.

Learning Objectives

1. Interaction issues and pitfalls: IRBs and investigators in the approval process
2. What kinds of information do IRBs need for initial and continuing review, and why?
3. When can an investigator use another IRB?
4. A special dimension of oversight: investigator fraud

Chapter 5 describes how an IRB operates, including sketches of some interactions with investigators, such as the kinds of communication that typically occur between investigators and the IRB. It is important to reiterate the basic way in which this occurs, as shown previously in Figure 5.1. This "paradigm" for both assuming and reporting responsibilities stresses the fundamental, "baseline" importance of the IRB/investigator flow of communication. The IRB comes closest to "touching" the subject only through the research site (where the investigator conducts research) and is able to discharge regulatory-mandated responsibilities only if it has access to appropriate information. Because some, if not most, of that information resides with the researcher or the sponsor, the IRB's access to the investigator or sponsor – without circumventing the investigator when a sponsor is involved – provides a critical communication path to assure appropriate oversight.

Under the best conditions, the investigator and the IRB can serve as mutual resources in the best interest of the subjects in a given study. Sadly, this is not always the case. Experienced investigators can narrate many tales of difficult interactions with IRBs; IRB staff, as well as IRB participants, in turn, can recount many instances when investigators have been particularly obstructive and obstreperous. Both these types of observation lead to the same interpretive outcome of inadequate protection of human participants in research.

INTERACTION ISSUES AND PITFALLS: IRBS AND INVESTIGATORS IN THE APPROVAL PROCESS

What are some of the typical vexing problems? Here are a few, taken from the investigator's perspective:

- The IRB's deadlines do not meet the reality of clinical research practice. Many IRBs meet at intervals that simply do not accommodate the time-lines of investigators or sponsors engaged in drug, device or biologics development. Given the potential to lose large sums of revenue (some conservative estimates use the loss of 1 million dollars per day that a new drug product is delayed in getting to market), sponsors find that many investigators and IRB's simply cannot remain competitive in the current research environment.
- The IRB does not appreciate how many constraints a research site is under, asking for things that add

administrative burdens without apparent benefit to the topic of research subject protection, especially given the fact that many investigators, particularly those in academic medical institutions, have responsibilities that are deemed to be directly related to the clinical care of patients and research subjects.

- Does the IRB not already have that information on file? After all, it has been the IRB of record for many of the same investigator's studies.
- The IRB does not seem to know what it is doing. How could the IRB possibly have deferred taking action on a project based on how informed consent will be obtained for research to be conducted in a critically ill population? Clearly, an IRB that has any concept of the potentially life saving therapies to be administered would not worry about the ethics of timely or adequate informed consent from those "patients."
- Calling the IRB for information causes more rather than less confusion. The IRB is not satisfied with a clarification from an investigator that a determination will be made to see if the patient is an appropriate candidate for investigational therapy from a clinical perspective. Why would the IRB be concerned about the "capacity" to give consent?

From the IRB's perspective, there are three important areas of concern.

- Incomplete information is submitted at the time of the initial review, causing numerous delays. Similarly, a lack of or incomplete information results in an additional administrative burden for IRB staff in maintaining the "tracking" of the current status of the research. Both concerns stretch the already understaffed capacity of the IRB.
- There is a failure to comply with regulations, despite reminders, for example with regard to reporting AEs and submitting continuing review information. This is an area of repeated IRB sanctions during audits, although, in fact, such activities are the regulatory (and ethical) responsibility of the investigator.
- Excessive delegation by the investigator to research and other support staff causes confusion for both staff and IRBs. Despite a formal commitment by the investigator to perform all study-related activities with the provision of

appropriate oversight, some investigators turn over most of the study activities to subinvestigators and study coordinators. These personnel may or may not be formally designated and/or authorized to participate in the research.

In addition, here are some typical pitfalls that investigators frequently encounter, whether or not they are within the investigator's control, and which trigger IRB concerns resulting in additional administrative or regulatory action.

1. *Overestimation/overoptimisim* In an attempt to remain competitive, some investigators promise to sponsors or funding agencies greater than feasible numbers of eligible subjects or shorter than feasible completion time-lines. While this overestimation in and of itself may not pose a threat to the protection of subjects' rights and welfare, such behavior may indicate other factors that compromise the ability of the investigator to provide appropriate and/or adequate oversight.
2. *Nonsponsor protocol modifications* There may be insistence on modifying the protocol for personal or subject benefit. The "bending" of certain criteria to accommodate the inclusion of certain patients (who would otherwise not be eligible) may expose subjects to additional risks that the IRB did not consider at the time of review and approval of the research. In addition, the contribution of an individual subject's data to the overall research may be invalidated, resulting in the unnecessary exposure of that specific person to risk.
3. *Study-related barriers* Taking on competing studies and delaying the start of one to complete another may result in greater financial or nonfinancial (e.g. prestige) gain.
4. *Administrative barriers* Examples are: "The CVs of the principal investigator and site personnel are in the process of being retyped and are therefore not available for submission to the IRB for review...;" or "The principal investigator's license is still good, even though it expired in 1997."
5. *Passing the buck* Examples are: "Paper-work has been submitted to the IRB; I don't know what the hold-up is!" or, "The sponsor was handling that serious AE we reported in a patient and I was told not to submit the information to the IRB..."
6. *Bailing-out* The canceling of or leaving study-related

meetings (investigator and other training sessions) early results in not receiving adequate research-related training.

7. *Deception* Data may be falsified.
8. *Misdispensing* "I know this is the tenth time but the pharmacist must have made a mistake."
9. *Unrandomizing* "I know subjects are supposed to be assigned to the treatment arms randomly; however, I know this patient and she really would be better off in the arm I put her in."
10. *Criteria Violations* "I thought the serum creatinine criteria seemed unrealistic for a patient with this condition and age." Again, such discretionary action on the part of the investigator places subjects at risk in situations previously not anticipated by the IRB at the time the research was reviewed and approved.
11. *Nonreporting of serious AEs or other AEs* "Hospitalization because of a stubbed toe is not a serious AE, is it?"
12. *Blind breaking* "I accidentally looked at the randomization assignment when the coffee spilled on the label and I used an alcohol swab to mop up the spills."

WHAT KIND OF INFORMATION DOES AN IRB NEED AND WHY?

An IRB collects information that is appropriate to its primary regulatory responsibility to protect the rights and welfare of human research subjects. It is charged with the responsibility of reviewing and approving research involving human subjects. The criteria by which an IRB evaluates "approvability" is a function of the disease or condition being studied in conjunction with safeguards proposed to address foreseeable risks and discomforts that a prospective human participant may encounter. These may range from physical to psychological to financial to societal to issues impacting on privacy and confidentiality.

As such, an IRB may request, in order to deliberate comprehensively, to review a range of information that it deems is material to its decision-making process for a given project. In today's environment, where an IRB may interpret the scope of its oversight role and authority, topics for consideration may include financial and other areas of conflict of interest; recruitment campaigns; investigator, support staff and/or site personnel core competencies; and the safeguards necessary for protection of the rights and welfare of research participants' data (such as administrative issues surrounding the management of technology). Important information is also collected regarding the content of the informed consent document and the process by which consent will be obtained.

Here are some additional examples of data that are often collected by IRBs and why they are important for human research subject protection:

1. *How many sites will the investigator oversee?* Is there appropriate personal oversight of the staff and appropriate conduct of the study in addition to the investigator simply being available to address any participant concerns (behavioral, clinical or otherwise). This issue is similar to the concern noted above regarding the investigator assuming more studies than can be conducted satisfactorily. One concern encountered in clinical research is whether the geographic distance between locations is routine and customary for that investigator's regular clinical practice.
2. *The ethnic composition of the investigation site's patients* There should be no potential special population or community issues. If there are, how are they to be addressed with specific regard to the research being proposed? Are there special considerations of the population being studied that require the imposition of additional safeguards specific to that group?
3. *If conducting clinical research, the medical license of the investigator* The investigator must be professionally and validly credentialed. While this is not a measure of the specific competency assessment of a clinical investigator, this document, in conjunction with others, serves to help to establish overall capability.
4. *Who obtains informed consent?* Is there a defined process for doing so? Because this is pivotal to the subject's understanding of the research and an area that could be used for both undue influence and coercion, the IRB reviews both who obtains informed consent and how it is obtained.

Regarding investigator responsibilities, the expectations are broadening at the present time. For instance, while the release of federally funded research is contingent on the "training of investigators," neither documentation of the scope and content of such training nor details of the comprehension and core competency of the investigator are required. A similar requirement for research involving FDA regulated products is not currently in place.

In the absence of standardized and acceptable guidelines, the requirements of investigators as noted on the FDA Form 1572 (used in the submission of drug research data to the FDA) – whether or not the investigator is conducting studies under the purview of the FDA – neatly describes criteria that all investigators can follow. Here are the critical portions from the "Statement of investigator," Sections 9 and 10 (regulations cited govern FDA regulated products, although the principles are applicable across all research):

"SECTION 9 COMMITMENTS:

I agree to conduct the study(ies) in accordance with the relevant, current protocol(s) and will only make changes in a protocol after notifying the sponsor, except when necessary to protect the safety, rights, or welfare of subjects.

I agree to personally conduct or supervise the described investigation(s).

I agree to inform any patients, or any persons used as controls, that the drugs are being used for investigational purposes and I will ensure that the requirements relating to obtaining informed consent in 21 CFR Part 50 and institutional review board (IRB) and approval in 21 CFR Part 56 are met.

I agree to report to the sponsor adverse experiences that occur in the course of the investigation(s) in accordance with 21 CFR 312.64.

I have read and understand the information in the investigator's brochure, including the potential risks and side effects of the drug.

I agree to ensure that all associates, colleagues, and employees assisting in the conduct of the study(-ies) are informed about their obligations in meeting the above commitments.

I agree to maintain adequate and accurate records in accordance with 21 CFR 312.62 and to make those records available for inspection in accordance with 21 CFR 312.68.

> I will ensure that an IRB that complies with the requirements of 21 CFR Part 56 will be responsible for the initial and continuing review and approval of the clinical investigation. I also agree to promptly report to the IRB all changes in the research activity and all unanticipated problems involving risks to human subjects or others. Additionally, I will not make any changes in the research without IRB approval, except where necessary to eliminate apparent immediate hazards to human subjects.
>
> I agree to comply with all other requirements regarding the obligations of clinical investigators and all other pertinent requirements in 21 CFR Part 312."
>
> "SECTION 10 SIGNATURE OF INVESTIGATOR:
>
> **(WARNING: A willfully false statement is a criminal offense. U.S.C. Title 18, Sec. 1001.)"**

A copy of the FDA Form 1572 or a similar testimony by the investigator to conduct research in a manner described resembling the content of 1572, especially the two sections just cited, is also often required by the IRB. A review of the "Commitments" section provides ample justification.

CONTINUING REVIEW: AN ONGOING OBLIGATION TO PROTECT RESEARCH SUBJECTS

A logical expectation of its involvement is for the IRB to provide continuing and ongoing oversight of issues impacting the protection of the rights and welfare of subjects participating in research. The spirit and intent of the regulations support this:

> "An IRB shall conduct continuing review of research covered by these regulations at intervals appropriate to the degree of risk, but not less than once per year, and shall have authority to observe or have a third party observe the consent process and the research" (21 CFR 56.109f; 45 CFR 46.109e)."

An area of current debate is how an IRB implements this mandate to fulfill the intent of providing the necessary continuing review.

While the regulatory expectation is that an IRB provides a review process similar to that performed initially when the research was approved, in practice how an IRB provides continuing review ranges from an in-depth re-review of research at defined intervals to a cursory evaluation of minimal information such as numbers of subjects enrolled and the experience obtained with those subjects. The presumption of the IRB in conducting the latter is that there is continuous monitoring of the research (such as the evaluation of serious AEs, modifications to the research, etc.) that justifies less stringent evaluation at those formal points in time when the documentation of re-review is necessary.

WHEN CAN AN INVESTIGATOR USE ANOTHER IRB?

In the traditional context of IRB involvement, an investigator conducting research in an institution is obligated to comply with the requirements of the IRB of that institution. The "decentralization" of research utilizing investigators in noninstitutional settings, particularly in drug, device or biologic development research for regulatory approval, has led to the need for IRB oversight that is nontraditional. Investigators who do not have access to an institution-based IRB or those whose IRBs simply cannot accommodate the time and administrative demands required, profile the users of independent IRBs.

When an investigator who is affiliated with an institution that has an IRB chooses to use another IRB, there should be appropriate documentation to indicate which IRB will have jurisdiction. There are different circumstances in which this may occur, but from a regulatory point of view the bottom line is clear: one IRB reviews and oversees the study at a particular site from start to finish, thereby insuring continuity in human research subject protection.

A SPECIAL DIMENSION OF OVERSIGHT: INVESTIGATOR FRAUD

Understanding the Issue

The Association of Clinical Research Professionals conducted a survey of its membership to support its White Paper presentation at their 2001 National Convention. Part of that survey addressed the public's opinion of clinical research, including issues of trust and credibility in the light of increased press coverage of international as well as domestic trials. A previous survey by the same group, conducted in 2000, revealed that the membership did not believe that widespread ethical lapses existed in clinical research, including the generation of fraudulent data. Sadly, the misbehavior of some investigators continues to plague all researchers.

In late 2001 the NBAC released its final report entitled *Ethical and Policy Issues in the Oversight of Human Research*. The report included NBAC's recommendations for overhauling the current system (NBAC, 2001). While fraud *per se* was not addressed in depth, the three general themes of their recommendations are worth noting as a "surrogate marker" of the commission members' appreciation of fraud prevention:

- "Our recommendations related to policy generally suggest that there should be fewer regulations and more guidance."
- "The recommendations...focus attention on research in which participants need to be protected, striving to make the level of protection commensurate with the level of risk."
- "Our recommendations somewhat increase the scope of regulated research but streamline the process of regulatory compliance."

From these kinds of recommendations one might infer that the Commissioners would relate actual or potential fraudulent behavior to the cumbersomeness of conducting the research, including defining how administratively taxing minimal risk research should be (a topic addressed at length in their Chapter 3).

More bluntly, the Office of the Inspector General's report on research subject recruitment and clinical investigators (OIG, 2000a,b) portrayed the atmosphere of clinical research as one of being highly pressurized and susceptible to fraud.

Practically speaking, no IRB should be put in the position of policing, in the name of research subject protection, the scientific misconduct of investigators, including fraud. The view from the IRB (that is, the vantage of ethical principles applied to human participant protection matters) incorporates the cumbersomeness of the clinical research process and the pressures that biomedical research enterprise face. Certainly, this is not to be so naïve as to dismiss a human factor such as greed. However, there does not appear to be a conviction among clinical research professionals that fraud is rampant in the system. Regardless of this, a practical approach to fraud is necessary, at least as a point of departure.

Fraud and the Cumbersomeness of Conducting Research

- What constitutes fraud?
- Is there such a thing as unintentional fraud (and does it differ from "negligence")?
- Is the most pressing problem of fraud largely one of creating false data or of fake record keeping?
- Do financial improprieties necessarily correlate with "dirty data"?
- How is fraud a human participant protection issue?

The IRB's perspective on the administrative burden of clinical trials is hardly unique. It does have an intimate double vantage: that of the IRB itself and that of the institution site. Regulatory and guidance requirements enhance the documentation process and the burden, not to mention the human resources needed to maintain a semblance of staying current.

The Office of Research Integrity, a sister agency to the OHRP within the same DHHS, has jurisdiction over issues of fraud and misconduct in science. Research with human subjects falls within the general category of research and is thus included in all processes of scientific misconduct policy. Through years of dialogue among major stakeholders on this issue, the federal government has eventually issued its final ruling on the definition of scientific misconduct and how it should be handled. The official definition of scientific misconduct (OSTP, 2000) is as follows:

> "Research misconduct is defined as fabrication, falsification, or plagiarism in proposing, performing, or reviewing research, or in reporting research results.
>
> Research, as used herein, includes all basic, applied, and demonstration research in all fields of science, engineering, and mathematics. This includes, but is not limited to, research in economics, education, linguistics, medicine, psychology, social sciences, statistics, and research involving human subjects or animals.
>
> Fabrication is making up data or results and recording or reporting them.
>
> Falsification is manipulating research materials, equipment, or processes, or changing or omitting data or results such that

the research is not accurately represented in the research record.

Plagiarism is the appropriation of another person's ideas, processes, results, or words without giving appropriate credit."

The process required to handle misconduct is that each research institution deals with the investigation and the final report is delivered to the federal agency. Each federal agency such as the NIH should have a detailed plan in place of the procedure to be followed. However, the federal rule prescribes that the federal agency has the right to conduct its own investigation to protect the public interest such as public health and safety. This is clearly meant to include the province of research involving human subjects (for more details see Shamoo and Resnik, 2002, in press).

As the NBAC report and earlier Office of the Inspector General's reports have noted, one of the major difficulties for IRBs is keeping up with the sheer volume of work. While laudable, calls for the accreditation and certification of IRBs, both staff and members, represent a yet greater burden, even if of a different kind, and no relief seems to be in sight for handling AEs both more effectively and efficiently. The net result can be hasty deliberations and carelessness in documentation and communication. In short, when a system is out of control, it invites fraud. This may well be considered fraud by omission as opposed to fraud by commission. Nevertheless, it results in breakdowns in human participant protection.

Whether this kind of fraud fulfills a legal definition or not, ignorance of the law, in this case of regulatory compliance, has not been upheld as a justification for sloppy work or inadequate oversight. Clearly, fraudulent data is one thing, but may fraud not exist in other ways, specifically, for example, by virtue of failures in documentation or inadequate quality assurance or quality control?

Although the gentler term "negligence" is more typically applied, the continuing wave of shutdowns of academic medical centers' IRBs in particular may suggest that something stronger is indeed appropriate.

On the research site side, the number of naïve investigators continues to increase due to the sheer volume of funding agencies, industry need, and turnover, coupled with the lack of qualified and trained investigators. In turn, this escalates communication tensions between IRBs and research sites. Attendant on increased communication is increased documentation, perpetuating the problem of administrative cumbersomeness and the potential for documentation error, leading to real or perceived fraudulent behavior.

From an IRB perspective, among experienced investigators intentional fraud involves the conscious submission of false information to an IRB, ranging from invalid medical licenses through choosing an IRB that is not duly constituted (a responsibility of the investigator as indicated on FDA Form 1572). Unintentional fraud in these situations may be exemplified by the inappropriate delegation of responsibilities to study staff, especially those who are unqualified.

To reiterate, the excessive administrative burdens of documentation and communication invite unintentional "fraud" via short cuts and omissions. The human factor of frustration, not simply greed, can lead directly to intentional as well as to unintentional fraud.

Fraud and the Pressures facing the Clinical Research Enterprise

There are two contradictory public sentiments that are vital to understanding the pressures facing the clinical research enterprise. The first is the public's desire for more and more cures quickly, with less and less risk. This is not to ignore the business and marketplace realities that add to the complexity and uncertainty related to the issues of fraud and possible misconduct.

One of the most compelling points in the NBAC (2001) report is the emphasis on the mutuality of taking responsibility, not only on the part of regulatory agencies, sponsors, investigators, funding agencies, and IRBs, but also on the part of the "human participants" (the Commission's favored term to replace "human subjects"). The Commission also notes that an atmosphere of "protectionism" exists, which may or may not be appropriate. Given the current political climate, this is the second public dichotomy: expecting the government to protect the individual more and more, while intruding less and less on personal decision making.

These contradictory public sentiments create a climate favorable to fraud. A point of discussion focuses on IRBs being involved in the gatekeeping role, for which they are neither equipped nor have the capability.

As human participant protection matters come more and more to the forefront of public discussion, the definition of fraud may well need to be expanded beyond the narrow confines of prosecutorial law, at least for ethical purposes. From a research participant perspective, terms like coercion denote both ethical as well as regulatory force.

Fraud: Towards an Ethical Decision-Making Model

The NBAC report underscores the importance of administrative matters by advising that "administrative review" should be used more widely to assist in overcoming the cumbersomeness that is part of the atmosphere that encourages fraud. In doing so, the Commissioners are also holding this type of review in serious regard, as the primary form of protection in cases of minimal risk.

Fraud may be viewed in terms of different "axes:" one represented by the extremes of intention and ignorance, another by those of conflicts of interest (actual or apparent), and a third by those of ethics and regulations. Making decisions about what is the appropriate thing to do needs to take all three into account.

1. *Intention/ignorance* What is the primary intention of the research professionals involved and is there adequate understanding that what is being done may be fraudulent?
2. *Conflicts of interest* Do the research professionals believe that the reason for potential concern (payments, stock ownership, etc.) will impact their work? How would the media describe the reason for potential concern? Would the research professionals feel comfortable submitting the information, whether regulatorily required or not, to an impartial third party?
3. *Ethics/regulations* What is the regulatory perspective on the matter of concern? What are any additional ethical considerations, especially with reference to basic ethical principles like beneficence, justice, and respect for persons?
4. *AE reporting* The system should define and put in place a clear policy and method of oversight for AE reporting. At present, the FDA AE reporting requirements are more clear and are adhered to than those agencies under the Common Rule (Shamoo, 2001).

Quiz: Choose the Best Response

1. Which of the following pieces of information should the IRB obtain:
   a. How many sites the investigator will oversee
   b. The ethnic composition of the site's patients
   c. The medical license of the investigator
   d. Who obtains informed consent
   e. All of the above

2. An investigator who conducts research with inpatients needs to submit information to:
   a. The hospital's IRB
   b. The IRB of the academic medical center at which the investigator is a member of staff
   c. The independent central IRB that is reviewing the study for the sponsor
   d. All of the above
   e. Only one of the above

3. Among the investigator's FDA Form 1572 responsibilities are:
   a. Oversight of staff
   b. Choice of IRB
   c. Maintaining adequate and accurate records
   d. All of the above
   e. Only a and c

4. There are federal regulations about:
   a. Obligations of investigators
   b. Fraud
   c. Initial and continuing review by an IRB
   d. a and c only
   e. a, b, and c

5. Fraud can be:
   a. An ethical matter
   b. A legal matter
   c. A regulatory matter
   d. All of the above
   e. Any combination of two of the above

Case Studies

Case 1

An investigator is conducting a study involving both inpatients and outpatients. The study begins in a hospital where the investigator has admitting privileges. When subjects are discharged they will continue in the study, being followed on an outpatient basis. The hospital is a large facility in an urban area and has an IRB as well as a clinical research unit serving a range of departments in the hospital that are participating in clinical trials.

From a regulatory and ethical perspective, how would you advise the investigator to go about making sure that there is appropriate IRB oversight?

Case 2

An investigator is conducting research within an academic medical center (AMC). She also has a private clinical practice where she conducts some industry studies. Because she is a well-known researcher in a particular field, she conducts both federally-funded as well as industry-sponsored research. Her AMC also conducts industry-sponsored research. In order to simplify a number of matters, she has separated her research within the AMC from that conducted in her private practice. Her government-sponsored research is conducted within the AMC; industry-sponsored research is conducted within her private practice.

What immediate or potential issues do you see with this approach? Are there areas of pressure that the IRB may want to be particularly sensitive to with respect to conflicts of interest?

Case 3

A naïve investigator (a formal term for a physician not experienced in clinical research) has a thriving private practice and decides to enhance this by conducting clinical research. Gearing up to do so, the investigator hires an experienced study coordinator and plans to bring in another physician to cover patients for him while he allots time to the research. Unfortunately, finding an appropriate physician becomes a difficult matter and the new investigator has landed two studies to conduct almost immediately. Being something of a workaholic and trusting the competency of the study coordinator, he not only begins both studies but is also in the process of committing himself to a third. The study coordinator's mother passes away and the coordinator must attend to her funeral and estate, necessitating being out of state for an extended period of time.

Into how many "pitfalls" can you imagine that the investigator may be likely to fall? Assuming you were called in to advise him, what suggestions would you make?

References

Food and Drug Administration (FDA) (1998) Information Sheets. Guidance for Institutional Review Boards and Clinical Investigators. Rockville, Maryland: FDA.

21 Code of Federal Regulations parts 50, 54, 56, 312, 314

National Bioethics Advisory Commission (NBAC) (2000) Ethical and Policy Issues in Research Involving Human Participation. Rockville, Maryland: NBAC. Available at: http://www.bioethics.gov

Office of the Inspector General (2000a) Recruiting Human Subjects: Pressures in Industry-Sponsored Clinical Research. Available at: http://www.dhhs.gov/progorg/oei

Office of the Inspector General (2000b) Recruiting Human Subjects: Sample Guidelines for Practice. Available at: http://www.dhhs.gov/progorg/oei

Office of Science and Technology (OSTP) (2000) Federal Register, vol. 65, 235. Available at: http://frwebgate.access.gpo.gov/cig

Shamoo AE (2001) Adverse events reporting – the tip of an iceberg. *Accountability Res.* 8, 197–218.

Shamoo AE & Resnik DB (2002) Textbook on Responsible Conduct of Research. New York: Oxford University Press (in press).

Answers

Chapter 1

1. d
2. e
3. e
4. a
5. b
6. b
7. d
8. b
9. e
10. c
11. e

Chapter 2

1. e
2. b
3. e
4. b
5. c
6. d
7. e

Chapter 3

1. e
2. b
3. d
4. b
5. e
6. d
7. e
8. e

Chapter 4

1. e
2. d
3. d
4. e
5. d

Chapter 5

1. c
2. c
3. e
4. d
5. d

Chapter 6

1. d
2. a
3. b
4. e
5. a

Chapter 7

1. e
2. c
3. b
4. a
5. d

Chapter 8

1. e
2. d
3. b
4. a
5. b
6. e
7. d

Chapter 10

1. e
2. e
3. d
4. e
5. d

Index

A

B

C

I

J

K

L

M

N

O

P

R

S